GENE TRANSFER METHODS:

INTRODUCING DNA INTO LIVING CELLS AND ORGANISMS

Other BioTechniques® Books Titles

Gene Cloning and Analysis by RT-PCR
P.D. Siebert and J.W. Larrick (Eds.)

Apoptosis Detection and Assay Methods
L. Zhu and J. Chun (Eds.)

Protein Staining and Identification Techniques
R.C. Allen and B. Budowle

Immunological Reagents and Solutions: A Laboratory Handbook
B.B. Damaj

Affinity and Immunoaffinity Purification Techniques
T.M. Phillips and B.F. Dickens

Yeast Hybrid Technologies
L. Zhu and G. Hannon (Eds.)

Viral Vectors: Basic Science and Gene Therapy
A. Cid-Arregui and A. García Carrancá (Eds.)

Ribozyme Biochemistry and Biotechnology
G. Krupp and R.K. Gaur (Eds.)

Antigen Retrieval Techniques: Immunohistochemistry and Molecular Morphology
S.-R. Shi, J. Gu, and C.R. Taylor (Eds.)

Bioinformatics: A Biologist's Guide to Biocomputing and the Internet
S. Brown

SNP and Microsatellite Genotyping: Markers for Genetic Analysis
A. Hajeer, J. Worthington, and S. John (Eds.)

Gene Transfer Methods:
Introducing DNA Into Living Cells and Organisms

Edited by

Pamela A. Norton and Laura F. Steel
Department of Biochemistry and Molecular Pharmacology
Jefferson Center for Biomedical Research
Thomas Jefferson University
Philadelphia, PA, USA

A BioTechniques® Books Publication
Eaton Publishing

Pamela A. Norton, PhD and Laura F. Steel, PhD
Department of Biochemistry and Molecular Pharmacology
Jefferson Center for Biomedical Research
Thomas Jefferson University
700 East Butler Avenue
Doylestown, PA 18901

Library of Congress Cataloging-in-Publication Data

Gene transfer methods : introducing DNA into living cells and organisms / edited by
Pamela A. Norton and Laura F. Steel.
    p. cm.
  Includes bibliographical references and index.
  ISBN 1-881299-34-1 (hardcover)
   1. Genetic engineering--Laboratory manuals. 2. Genetic transformation--Laboratory
manuals. I. Norton, Pamela A. II. Steel, Laura F.

QH442.6 .G45 2000
660.6'5--dc21
                                                00-062314

ISBN 1-881299-34-1

Printed in the United States of America

9 8 7 6 5 4 3 2 1

Eaton Publishing
BioTechniques Books Division
154 E. Central Street
Natick, MA 01760
www.BioTechniques.com

Francis W. Eaton: *Publisher and President*
Stephen Weaver: *Director and Editor-in-Chief*
Christine McAndrews: *Managing Editor*
Sandy Lamont: *Production Manager*
Jayne Segman: *Cover Designer*

CONTRIBUTORS

Sanford I. Bernstein
Department of Biology and
Molecular Biology Institute
San Diego State University
San Diego, CA, USA
(sbernst@sunstroke.sdsu.edu)

Andrew N. Binns
Department of Biology
Plant Science Institute
University of Pennsylvania
Philadelphia, PA, USA

Richard M. Cripps
Department of Biology
University of New Mexico
Albuquerque, NM, USA

Elizabeth L. George
Vascular Research Division
Department of Pathology
Brigham and Women's Hospital and
Harvard Medical School
Boston, MA, USA
(egeorge@rics.bwh.harvard.edu)

R. Daniel Gietz
Department of Human Genetics
University of Manitoba
Winnipeg, Manitoba, Canada

Larry G. Johnson
Division of Pulmonary and
  Critical Care Medicine
Department of Medicine
The Cystic Fibrosis/Pulmonary
  Research and Treatment Center
The University of North Carolina at
  Chapel Hill
Chapel Hill, NC, USA
(Larry_Johnson@med.unc.edu)

David A. Knecht
Department of Molecular and Cell
  Biology
University of Connecticut
Storrs, CT, USA
(knecht@uconn.edu)

Eunkyung Lee
Department of Molecular and Cell
  Biology
University of Connecticut
Storrs, CT, USA

Stuart A. Newman
Department of Cell Biology and
  Anatomy
New York Medical College
Valhalla, NY, USA
(newman@nymc.edu)

Pamela A. Norton
Department of Biochemistry and
  Molecular Pharmacology
Jefferson Center for Biomedical
  Research
Thomas Jefferson University
Philadelphia, PA, USA
(Pamela.Norton@mail.tju.edu)

Carolyn Raffensperger
Science and Environmental Health
  Network
Windsor, ND, USA

Michael A. Savka
Department of Biological Sciences
Rochester Institute of Technology
Rochester, NY, USA
(massbi@rit.edu)

Laura F. Steel
Department of Biochemistry and
  Molecular Pharmacology
Jefferson Center for Biomedical
  Research
Thomas Jefferson University
Philadelphia, PA, USA
(Laura.Steel@mail.tju.edu)

Robin A. Woods
Department of Biology
University of Winnipeg
Winnipeg, Manitoba, Canada
(Robin.Woods@uwinnipeg.ca)

PREFACE

Few if any areas of biology and medicine remain untouched by the advances in molecular biology that have occurred over the last few decades. Tremendous progress has been made in our understanding of fundamental biological phenomena such as the control of cell division and the develoment of multicellular organisms. New therapeutic and diagnostic agents have been developed using molecular tricks to identify and co-opt genes and their products. Genetically modified organisms have begun to appear outside of the research setting. As we look ahead into the next century and the new millennium, it seems likely that technological advances will continue apace.

A critical aspect of the molecular revolution is that after experimental manipulation of DNA in vitro, the DNA must be introduced into living cells for it to manifest activity. At the most basic level, introduction into *Escherichia coli* is essential for the propagation of most recombinant DNA molecules. In addition, it is desirable to introduce DNA into cells of other organisms, ranging from yeast and fungi, to mammals and plants, or even to embryos and adults of many species. A variety of methods have been developed to accomplish these diverse tasks, and protocols for individual applications can be found in the primary literature as well as in collections of protocols in cell and molecular biology. The goal of this volume is to compile a core set of methods in a single volume, and contributions have been solicited from experts in the art of gene transfer. As the choice of method typically is dictated by the target organism, the volume is organized in this fashion. However, there is necessarily some overlap in topics where appropriate.

In addition to the presentation of detailed protocols, each chapter includes information regarding critical aspects of the method. Common problems are described along with specific troubleshooting suggestions. To augment the information presented, the reader is directed to the pertinent published literature, with an emphasis on primary sources. Where appropriate, mechanisms of DNA transfer are discussed, including the intracellular fate of newly introduced DNA. Discussion of alternative methodologies is presented to allow the reader some perspective on the field. Finally, some of the societal implications of the use of gene transfer technology are considered. We hope that this work will prove valuable to novice workers needing experimental guidance as well as experienced researchers seeking additional insight into familiar topics.

Pamela A. Norton
Laura F. Steel
May 2000

Contents

5 INTRODUCTION OF DNA INTO CULTURED MAMMALIAN CELLS . . 65
P.A. Norton

1 An Introduction to Gene Transfer

Pamela A. Norton[1] and Laura F. Steel[2]
[1]*Division of Gastroenterology and Hepatology, Department of Medicine, and* [2]*Department of Biochemistry and Molecular Pharmacology, Jefferson Center for Biomedical Research, Thomas Jefferson University, Philadelphia, PA, USA*

The first gene transfer experiment was reported in 1928 by Griffith (2), who observed that heat-killed pathogenic pneumococcus, when mixed with cells of a nonpathogenic strain, could transform the latter to become pathogenic. However, the transforming substance was not identified in these experiments, and another 16 years elapsed before Avery and colleagues (1) demonstrated carefully that deoxyribonucleic acid, or DNA, was the transforming activity. Subsequently, Hershey and Chase (3) showed that DNA was the only material transferred during bacteriophage infection, suggesting the generality of the finding that nucleic acids are the genetic material. These pioneering studies, followed by those that elucidated the structure of DNA, the basis of its replication, and the nature of the information encoded in the molecule, laid the groundwork for a profound change in the biological sciences.

Other important developments, although somewhat less visible, have in fact proven equally important to the molecular revolution. Specifically, deliberate modification of DNA molecules is not an end in itself; the DNA must be re-introduced into a living cell or organism in order to study the functional consequences of the DNA alteration. Such reverse genetics relies on the existence of methods to accomplish the introduction of manipulated genes. The earliest DNA transfer experiments were the same ones which demonstrated that DNA was the genetic material (1); the process was dubbed transformation, as the bacteria were transformed from one phenotype to another. Although bacteria will take up DNA without any special treatment, the efficiency of the process is very low. Subsequently, the efficiency of DNA transfer into bacteria was enhanced with the discovery that the addition of divalent cations facilitated DNA uptake by *Escherichia coli* (4). Numerous refinements to this basic protocol have been devised, but the simple method of $CaCl_2$-

Genes Transfer Methods
Edited by Pamela A. Norton and Laura F. Steel
©2000 Eaton Publishing, Natick, MA

based transformation remains in use to this date as described in Chapter 2. This chapter also includes methods for the purification of plasmid DNA from *E. coli*, as this process is essential for most of the procedures in the subsequent chapters.

Reverse genetics has neatly complemented conventional genetic and biochemical strategies in ascertaining the function of individual genes. Modern research into fundamental biological mechanisms has tended to focus on a small number of key model organisms. Much of our understanding of these mechanisms has derived from studies of simple organisms such as yeast and fungi, which are well-suited to experimental manipulation. A review of several commonly used methods to introduce DNA into the important experimental organism *Saccharomyces cerevisiae* is presented by Woods and Gietz in Chapter 3. Another simple eukaryotic organism, *Dictyostelium discoideum*, is often used as a model for cell differentiation and locomotion. In Chapter 4, Knecht and Lee present detailed methods for introducing DNA into *Dictyostelium* and describe an unusual method to enhance targeted mutagenesis. The ongoing accumulation of complete genomic sequences will create a need for methods to transfer DNA into an ever expanding number of organisms.

The basic techniques for introducing DNA into *E. coli* have inspired procedures for the introduction of DNA into cells from a wide variety of organisms, including vertebrates. Early work on DNA transfer into mammalian cells stemmed from studies of DNA tumor viruses; cationic polymers or calcium phosphate co-precipitation were shown to improve the very inefficient uptake of naked DNA. Subsequently, these basic protocols have been refined, and many new alternatives have been developed, including liposome-mediated transfer and electroporation, as discussed in Chapter 5. Transfection efficiencies of nearly 100% have been reported for the right combination of methodology and cell type.

An exciting prospect for geneticists and developmental biologists has been the ability to transfer genes into the germ lines of metazoans such as flies, mice, and plants. It took many years before gene transfer methodology was developed for the intensively studied dipteran *Drosophila melanogaster*. Success relied on the use of a transposable element called the P element. As detailed by Cripps and Bernstein in Chapter 6, insertion of the desired gene into a cloned P element and the introduction of DNA into an embryo of appropriate genotype will lead to transposition of the DNA into one of the host chromosomes. Resulting individuals are chimeric for the introduced DNA, but successful transfer into cells of the germline results in progeny flies who will transmit the gene to subsequent generations in a Mendelian fashion.

An even simpler strategy proved highly successful for gene transfer into the germline of the best genetically characterized mammal, the mouse. DNA is microinjected into the pronuclei of fertilized one-cell mouse embryos, which are then transferred back to foster mother mice. The DNA integrates into the genome at random sites, and generally the new transgene is transmitted to the next generation with high frequency. The subsequent development of techniques to permit site-specific integration of DNA was an exciting advance. As presented by George and Milstone in Chapter 7, the desired DNA is introduced into totipotent embryonic stem cells. The cells containing the desired genetic constitution are identified and re-introduced into blastocysts, which are then transferred to the uteri of female mice. This methodology has most commonly been used to inactivate genes by insertional mutagenesis; the technology is continually being refined so as to produce ever more precise genetic alterations.

Introduction of DNA into plants is of great agricultural potential and medical importance, in addition to its value as a research tool to understand basic plant biology. Several techniques, described by Savka and Binns in Chapter 8, have been employed to transfer DNA into plants, either directly or by transfer into leaf fragments, which are then used to regenerate intact plants. Much work has relied on the use the bacterium *Agrobacterium tumorfaciens* as a highly efficient vector for gene delivery to some plant species. The use of biolistic technology to introduce naked DNA directly into plant cells is expanding the range of plant species that can be transformed.

Research in gene transfer methodologies and uses has inevitably brought us to where it is now possible to alter the human genome. This technology has far-ranging applications in the treatment of both genetic and acquired diseases. At present, most gene therapy strategies are designed to introduce DNA into the affected somatic cells without modifying the germline, as such alterations pose substantial ethical dilemmas. Not surprisingly, the application of gene transfer methods to humans is vastly more complicated, even at the technical level, than for other organisms, and thus, it is not possible to present specific protocols. However, in Chapter 9, Johnson discusses in detail both the promise of these methods and the barriers that must be overcome in transferring genes to cells of the airway, a tissue that has received considerable attention as a therapeutic target. Although the obstacles remain formidable, it is likely that gene therapy will become a reality for the treatment of at least some human diseases.

Most of the methods presented in this volume are designed with laboratory research applications in mind. After providing detailed how to protocols for gene transfer, Chapter 10 presents an essay that urges great caution in their application. In elaborating the Precautionary Principle, Newman and Raffensperger challenge us to consider the potential significance of various applications of gene transfer methods to society at large. In some cases, the authors would have scientists stop gene transfer experiments altogether until there has been much more study of potential adverse outcomes. These views will be seen as appropriate by some and as extreme and unrealistic by others. Nevertheless, as the applications of gene transfer experiments increasingly manifest themselves in everyday life, the debate about the merits and risks of the technology is moving into the public arena. It will be incumbent upon the practitioners of the art and science of gene transfer to be engaged in this discussion. We hope that the presentation of these ideas will stimulate debate among scientists and lead to increased communication regarding the risks and benefits of new technologies to the public.

REFERENCES

1.Avery, O.T., C.M. MacLeod, and M. McCarty. 1944. Studies on the chemical nature of the substance inducing transformation of pneumococcal types. J. Exp. Med. *79*:137-158.
2.Griffith, F. 1928. The significance of pneumoccocal types. J. Hyg. *27*:113.
3.Hershey, A.D. and M. Chase. 1952. Independent functions of viral protein and nucleic acid in growth of bacteriophage. J. Gen Physiol. *36*:39-56.
4.Mandel, M. and A. Higa. 1970. Calcium dependent bacteriophage DNA infection. J. Mol. Biol. *53*:19-162.

2 | Transformation of Bacteria and Purification of Plasmid DNA

Laura F. Steel
Department of Biochemistry and Molecular Pharmacology, Jefferson Center for Biomedical Research, Thomas Jefferson University, Philadelphia, PA

INTRODUCTION

Projects in which new genes will be introduced into eukaryotic cells or organisms very often involve the preparation of those genes, or gene segments, by cloning into plasmid vectors with subsequent propagation in and purification from bacterial hosts. Specific strategies for gene design and construction, and the choice of appropriate host-vector systems, will vary with each application and are discussed in individual chapters. Nevertheless, general methods for efficient bacterial transformation and the reliable preparation of highly purified plasmid DNA are fundamental to the success of all gene transfer experiments and are presented in this chapter.

BACTERIAL TRANSFORMATION

Chemical and Electrical Methods of Bacterial Transformation

The process of transformation requires that a relatively large, electrically charged DNA molecule cross the barrier presented by the bacterial cell envelope to establish itself for expression and replication within the cell. Without experimental manipulation of the cells and DNA, this is a very rare event. However, early studies by Mandel and Higa (11) showed that the efficiency of uptake of purified bacteriophage DNA could be greatly increased if bacterial cells (*Escherichia coli*) were incubated in the presence of cold calcium choride, mixed

Gene Transfer Methods
Edited by Pamela A. Norton and Laura F. Steel
©2000 Eaton Publishing, Natick, MA

with the DNA, and then subjected to a brief heat treatment at 37°C. These observations were extended when it was shown that the state of competence induced by treatment with cold calcium chloride also allows uptake of other types of DNA, including fragments of chromosomal DNA (12) and plasmids (4). The precise mechanism of DNA entry into bacterial cells during transformation remains unexplained, but it is generally thought that divalent cations such as calcium help to organize the interaction between DNA molecules and the bacterial cell surface, both of which are negatively charged. Maintenence of the cells at low temperature, with a brief heat shock at 37° to 45°C, is thought to promote DNA uptake by causing phase transitions in the phospholipid-containing membranes of the cell envelope (see discussion in Reference 7).

Many investigators have explored modifications to the early procedures to increase the efficiency of transformation (the number of transformed cells/µg DNA). Improvements to the procedure have been based primarily on empirical observations made during systematic evaluations of various parameters related to the preparation of competent cells and the transformation itself. For instance, several groups have investigated the effects of cell growth conditions [growth medium components (7), temperature (9), and cell density at time of harvest (7,9,16)], di- and mono-valent cation combinations used to induce competence (7,9,10), additions such as dimethyl sulfoxide (DMSO) or dithiothreitol (DTT) to the transformation mixture (7,10), and the duration of cold and heat incubations (5,9). Conclusions from different groups are not always in agreement, and optimized procedures can be highly species and strain specific. Nevertheless, competent cells from commonly used cloning strains of *E. coli* are routinely prepared and transformed by any of several similar procedures.

Two procedures are presented below. The first is essentially that of Mandel and Higa (11), and is designed to be quick and easy. It will yield cells with a transformation efficiency of about 10^6 transformants per microgram of supercoiled DNA, which is often entirely adequate. The second procedure is based on one developed by Hanahan (7) after a comprehensive investigation of transformation parameters. It will produce cells of considerably higher transformation efficiency ($>10^7-10^9$ transformants/µg supercoiled DNA), that can be stored frozen while maintaining that level of efficiency. Other procedures have been reported (e.g., see Reference 9) that may produce even higher efficiencies.

Bacterial cells can also be transformed by electroporation (6,15). Just as for transformation of chemically treated cells, the mechanism of DNA uptake by electroporation is not fully understood. Exposure to a strong electrical pulse causes a transient increase in the permeability of cells that allows DNA to cross the cell envelope, probably through channels created by the pulse. The efficiency of transformation by this method increases both with the strength of the electrical field and the time of the pulse. Considerably higher field strengths are used for electroporation of bacterial cells than mammalian cells, and a special apparatus is used that includes a circuit to control the pulse, protecting the circuit and preventing arcing (6). Ultimately, the efficiency of transformation is limited by the tradeoff between an increase in the number of cells that take up DNA and a decrease in cell viability as either the electric field strength or the pulse length is increased. However, it is possible to achieve transformation efficiencies of $>10^9$ to 10^{10} transformants per microgram of DNA, where only 25% to 50% of the cells survive, but more than 80% of the surviving cells are transformed (6). The preparation of cells for electroporation simply involves growing them to mid-log phase and then washing several times in water to reduce the ionic strength and concentrate the cell suspension. Electrocompetent cells can be stored frozen for later use.

As compared with chemical methods, transformation by electroporation is linear over a wider range of DNA concentration, results in higher efficiencies, and is less subject to variability based on the recipient strain, the size of the plasmid, or the form of the DNA (supercoiled or ligated). However, it is also more expensive because of the requirement for a specialized apparatus and single-use electroporation cuvettes.

Maximum transformation efficiency is not required when the goal is propagation of a previously constructed plasmid. Cost, convenience, and speed of the procedure are likely to be more important considerations. However, both convenience and reasonably high efficiency can be achieved by preparing a large batch of competent cells that can be stored frozen and used as needed over a period of many months. While procedures for preparing highly competent cells are relatively easy and inexpensive (see protocols below), some laboratories, for instance those that will only need to do a small number of transformations, will find it advantageous to buy frozen competent cells. These are available from several manufacturers and are provided in different grades that will transform with varying levels of efficiency. Both chemically competent cells, with transformation efficiencies ranging from $>10^6$ to $>10^9$ transformants per microgram of supercoiled DNA, and electrocompetent cells, with an efficiency of $>10^{10}$ transformants per microgram of DNA, are commercially available.

Protocols

Preparation of Competent Cells

<u>Rapid Preparation of Competent Cells [based on Mandel and Higa (11)]</u>

This is a rapid and simple method for preparing competent cells that can be transformed with an efficiency of 10^6 transformants per microgram of DNA or higher. The competent cells can be stored at -80°C, but transformation efficiency is likely to decrease.

1. Inoculate 2 mL of LB broth with cells picked from a single colony of cells, or scraped from a tube of frozen cells, and grow overnight at 37°C with shaking.

2. Use 100 μL of the overnight culture to inoculate 50 mL LB in a 250-mL flask. Grow cells to mid-log phase (OD_{600} of approximately 0.5), for about 3 hours at 37°C with shaking.

3. Pellet the cells by centrifugation for 5 minutes at 4°C and 3000× g (e.g., 5000 rpm in an SS-34 rotor [Sorvall, Newtown, CT, USA]). For 50 mL of cells, split to two tubes for centrifugation.

4. Resuspend the pelleted cells in 25 mL $CaCl_2$/piperazine-N,N'-bis[2-ethanesulfonic acid] (PIPES) solution (1/2 volume) that is pre-chilled and ice cold.

5. Incubate on ice for 25 minutes.

6. Centrifuge cells at 3000× g for 5 minutes at 4°C.

7. Gently resuspend the pelleted cells in 5 mL ice cold $CaCl_2$/PIPES solution (1/10 original volume). Keep cells on ice until use. Cells can be used for at least 24 hours after preparation (some strains increase in competency during this time).

Notes: *(i)* PIPES is optional in $CaCl_2$/PIPES, but it helps to maintain the pH of the solution if stored for long periods. (*ii*) It is important to keep the cells ice cold throughout the procedure and to treat them fairly gently after exposure to $CaCl_2$. (*iii*) Competent cells can be used for 24 hours or longer after preparation and some strains will increase in competency during a day of storage on ice (5). (*iv*) If cells are to be stored frozen, include 10% glycerol in the $CaCl_2$/PIPES solution for final resuspension of the cells. Aliquot the cells, freeze quickly on dry ice, and store at -80°C.

<u>Preparation of High Efficiency Competent Cells [based on Hanahan (7)]</u>

Although this procedure is somewhat more involved than that given above, it results in cells that can be transformed at higher efficiency (greater than 10^8 transformants/μg DNA) and that survive storage at -80°C longer and more consistently. This procedure is recommended for laboratories that plan to do bacterial transformations routinely and will maintain a supply of frozen competent cells.

1. Inoculate 2 mL of SOB broth with cells picked from a single colony of cells, or scraped from a tube of frozen cells, and grow overnight at 37°C with shaking and good aeration.

2. Use 1 mL of the overnight culture to inoculate 30 mL SOB that has

been prewarmed to 37°C in a 500-mL flask. Shake at 37°C for 1 1/2 hours.

3. Use 1 mL from the growing culture to inoculate a second flask containing 30 mL prewarmed SOB. Incubate with shaking for 2 hours at 37°C.

4. Place the cells in a screw-cap centrifuge tube and incubate on ice for 10 minutes.

5. Pellet the cells by centrifugation at 1000× g (e.g., 3000 rpm in an SS-34 rotor) for 10 minutes at 4°C.

6. Resuspend the cells in 10 mL ice cold FSB/glycerol and incubate on ice for 10 minutes.

7. Pellet the cells by centrifugation at 1000× g for 10 minutes at 4°C.

8. Gently resuspend the cells in 2.4 mL FSB/glycerol, keeping the cells ice cold. Incubate on ice for 10 minutes.

9. Add 84 μL DMSO and incubate on ice for 5 minutes.

10. Add another 84 μL DMSO and incubate on ice for 5 minutes.

11. Cells can be used directly or stored at -80°C in aliquots (e.g., 100 μL) quickly frozen on dry ice or in liquid nitrogen.

Notes: (*i*) Reagents, including water, should be of high purity. (*ii*) Oxidation products of DMSO may be inhibitory (but see Reference 9). A fresh bottle of DMSO can be aliquoted and stored in microfuge tubes at -20°C. The unused portion of an aliquot should be discarded after a single use. (*iii*) It is a good idea to determine the transformation efficiency of each new batch of competent cells using a serial dilution of supercoiled plasmid DNA of known concentration. The cells should maintain this level of efficiency during storage for many months.

<u>Preparation of Cells for Electroporation [based on Dower et al. (6)]</u>

1. Inoculate 10 mL LB with cells picked from a single colony and grow overnight with shaking and good aeration at 37°C.

2. Use 2.5 mL overnight culture to inoculate 500 mL LB in a 2-L flask. Grow cells with shaking at 37°C to mid-log phase, an OD$_{600}$ of about 0.6.

3. Chill the cells on ice for 10 minutes, then harvest them by centrifugation at 4000× g [e.g., 5000 rpm in a GSA rotor (Sorvall), where cells will have to be split into several bottles] at 4°C for 10 minutes.

4. Remove the supernatant, drain well, and resuspend the cells in a few

milliliters of ice cold water. When the cells are evenly suspended, bring the total volume to 500 mL ice cold water.

5. Pellet the cells by centrifugation at 4000× g at 4°C for 10 minutes.

6. Pour off the supernatant, taking care not to lose cells from the loose pellet. Resuspend the pellets in a total volume of 250 mL water.

7. Pellet the cells by centrifugation at 4000× g at 4°C for 10 minutes. Remove the supernatant. Resuspend the cells in a total volume of 20 mL ice cold 10% glycerol. Combine cells and transfer to a single, smaller, prechilled centrifuge tube.

8. Pellet the cells by centrifugation at 3000× g (e.g., 5000 rpm in an SS-34 rotor) for 10 minutes at 4°C. Remove the supernatant.

9. Resuspend the cells by adding a volume of ice cold 10% glycerol that is about equal to the volume of the pellet (approximately 1.0 mL). The cell concentration should be at least 2 to 4×10^{10}/mL.

10. Aliquot the cells, freeze on dry ice, and store at -80°C.

Notes: (*i*) If 1 mM HEPES, pH 7.0, is used in the initial washes instead of water, the cells will form a tighter pellet. (*ii*) If cells are to be used immediately, and not stored frozen, glycerol is not needed in the final washes. (*iii*) It is a good idea to determine the transformation efficiency of each new batch of competent cells using a serial dilution of supercoiled plasmid DNA of known concentration. The cells should maintain this level of efficiency during storage for many months.

Transformation of Chemically Competent Cells

1. Add 50 μL competent cells to a tube that has been prechilled on ice.
2. Add DNA (usually in Tris-EDTA [TE] or TE^{-4}) in a volume of less than 10 μL. Incubate the cells with DNA for 25 minutes on ice.
3. Transfer the tubes to a water bath and incubate at 41°C for exactly 90 seconds.
4. Place the cells back on ice briefly, and then add 0.25 mL SOC and incubate at 37°C for 45 minutes.
5. Plate the contents of each tube on a standard LB agar plate that contains antibiotic appropriate for selection of the plasmid that is being used. Incubate overnight at 37°C.

Notes: (*i*) Smaller amounts of cells, e.g., 30 μL, also work well. (*ii*) It is important to keep the cells ice cold throughout this procedure, until the heat shock step. (*iii*) The amount of DNA can be adjusted so that 50 to 500 colonies are expected on a standard 100-mm plate.

Electroporation

1. Keeping components ice cold, mix 20 to 25 μL electrocompetent cells with 1 μL plasmid DNA (up to 100 ng).

2. Add the cells and DNA to an electroporation cuvette that has been pre-chilled on ice.

3. Place the cuvette in an electroporator and pulse the cells using settings recommended by the manufacturer. For *E. coli*, settings should result in the delivery of a pulse of about 12.5 to 16 kV/cm for 5 to 7 milliseconds.

4. Remove cells from the cuvette immediately after the pulse and add them to 1 mL SOC medium at room temperature. Incubate at 37°C for 1 hour.

5. Dilute the cells as necessary with SOC (to produce up to approximately 500 colonies per plate) and plate 0.1 to 0.2 mL on an LB plate with antibiotic appropriate for selection of the plasmid. Incubate the plates at 37°C overnight.

Notes: (*i*) Cells must be kept ice cold at all steps up to and including the electrical pulse. Tubes, cuvettes, and the electroporation chamber should be prechilled on ice. If frozen electrocompetent cells are used, they should be kept cold while thawing. (*ii*) The specific conditions (including volume of electrocompetent cells and amount of DNA) for electroporation will depend on the equipment and cuvette size being used, so it is important to follow the manufacturer's recommendations. (*iii*) The resistance of the mixture of bacterial cells and DNA must be high for electroporation, and salts added with the DNA must be kept to a minimum. If the DNA is not in a low salt buffer such as TE or TE^{-4}, it should be diluted, or better, precipitated and resuspended to remove salts. (*iv*) Cell survival drops rapidly (decreasing the number of transformants recovered) if cells are left in water after the electrical pulse (6), so it is important to dilute them into SOC as quickly as possible.

PURIFICATION OF PLASMID DNA

General Methods for Plasmid Purification

The purification of plasmid DNA, whether on a preparative or analytical scale, begins with the preparation of a cleared lysate from a saturated culture of bacterial cells that carry the plasmid of interest. Cells are harvested from the culture by centrifugation and then suspended for lysis by any of several methods. For instance, cells can be lysed with detergents (3), by treatment with lysozyme followed by boiling (8), or by lysis with sodium dodecyl sulfate

(SDS) at an alkaline pH (1,2). The cell lysate is then subjected to a round of denaturation and renaturation and soluble plasmid DNA is separated from the insoluble material containing chromosomal DNA, protein, and some of the cellular RNA by centrifugation. Plasmid DNA can be further purified from the supernatant, or cleared lysate.

The alkaline lysis method (1,2) is used in the procedures presented in detail below. It is widely used and yields cleared lysates with plasmid DNA that is somewhat less contaminated with cellular RNA and proteins than that made by other methods. In this procedure, cells are suspended in a solution containing glucose/Tris-HCl/EDTA and then incubated briefly with lysozyme. The addition of SDS and NaOH lyses the cells and raises the pH to 12.0 to 12.5, denaturing the chromosomal DNA. The solution is neutralized by the addition of 3 M potassium acetate, pH 5.5, resulting in the formation of an insoluble network of partially renatured high molecular weight DNA. A major portion of the cellular protein and RNA is precipitated at this step by the high salt concentration and SDS. Plasmids are small circular DNAs that are not denatured at pH 12.0 to 12.5 and remain soluble in the lysate. Insoluble material is cleared from the lysate by centrifugation. The soluble plasmid DNA can then be recovered from the supernatant.

The degree to which further purification is needed will depend on how the plasmid DNA is to be used. Frequently, it is sufficient for analytical purposes simply to precipitate DNA from the cleared lysate. A moderate level of additional purification can be achieved by RNaseA treatment and/or phenol extraction. However, to prepare plasmid DNA that is suitable for consistently efficient eukaryotic cell transfection, it is necessary to be more diligent in removing contaminants. Two methods, each with its own advantages, are frequently used: (*i*) banding in cesium chloride-ethidium bromide (CsCl/EtBr) density gradients, and (*ii*) chromatography on anion exchange resins.

The purification of plasmid DNA by centrifugation in CsCl/EtBr density gradients is based on the differential binding capacity of various forms of DNA for the intercalating dye EtBr (3,13). Physical constraints limit the amount of EtBr that will intercalate in small supercoiled plasmids relative to the amount bound by linear DNA or bacterial chromosomal DNA. Since binding of EtBr causes a decrease in the buoyant density of DNA, plasmid DNA will remain more dense and band at a higher density than linear or bacterial chromosomal DNAs during centrifugation in CsCl gradients in the presence of EtBr.

Plasmid purification methods that use anion exchange take advantage of the strong affinity of negatively charged DNA molecules for positively charged diethylaminoethyl (DEAE) resins. Potential contaminants of DNA preparations, such as RNA, proteins, single-stranded DNAs, and polysaccharides bind with varying, but lower affinities than DNA. These molecules either flow through an anion exchange column without binding, or can be washed from the resin at moderate salt concentrations that allow DNA to remain bound. The DNA is then eluted in a high salt buffer. It is important to note that the separation of plasmid DNA from chromosomal DNA occurs during the preparation of the cleared lysate, as these DNA forms will elute together from the column.

Both CsCl/EtBr gradient centrifugation and anion exchange chromatography will produce highly purified DNA that is suitable for cellular transfections. However, in choosing a method, considerations such as DNA yield, ease of use, availability of equipment, and cost will be important.

Plasmid DNA purification by CsCl/EtBr gradient centrifugation remains the gold standard for comparison to other methods. A major advantage of this method, besides the purity of the product, is its potential for providing large amounts of DNA at relatively low cost (not counting the ultracentrifuge). For high copy number plasmids, such as those derived from the pUC plasmids, quantities in the range of several milligrams can be recovered from a few hundred milliliters of saturated cell culture. This can be beneficial when repeated transfections will be done using a particular recombinant plasmid, for instance as a control or cotransfection plasmid in a series of experiments.

Disadvantages of CsCl gradient-based methods of plasmid purification are the higher degree of skill and attention required for preparation and packaging of the DNA/CsCl/EtBr mixture in sealed centrifuge tubes, the organic extractions required for postgradient cleanup of the DNA, and the use of hazardous chemicals (ethidium bromide and organic solvents). In addition, the method can require up to two days to complete, although much of that time is hands-off time during centrifugation.

Anion exchange columns have become very popular for plasmid purification because they are relatively easy to use, and preparations can be completed in half a day, yielding high purity DNA without the use of hazardous chemicals. The maximum yield from these preparations is determined by the binding capacity of the column, with cost rising with capacity. A variety of columns that will yield from 20 μg to 10 mg of plasmid DNA are now available from several manufacturers.

The yield and purity of plasmid DNA in any preparation will also be affected by characteristics of the bacterial host cells and the plasmid vectors themselves. Commonly used host strains carry the mutation *endA1*, which improves plasmid yield by removing an endonuclease activity, and *recA*, which stabilizes the plasmid construct by reducing recombination. In addition, most contain a mutation at the hsdR locus that inactivates the restriction system so that both modified (methylated) and unmodified plasmids can be grown. Many strains are suitable, but problems can occur if, for instance, the host grows poorly or produces high levels of endonuclease or carbohydrates. Plasmid vectors that are chosen for the preparation of DNA for transfection will usually contain an origin that directs replication to high copy number in bacterial cells. The yield from low copy number plasmids (including those built in a backbone derived from pBR322) can be increased by amplification in the presence of chloramphenicol (see Reference 14). Procedures presented below are designed for high copy number plasmids, such as those derived from the pUC series of vectors.

Protocols

CsCl/EtBr Gradient Purification of Plasmid

An abbreviated version of this procedure is presented in Table 1.

<u>Prepare Cleared Lysate and Precipitate Plasmid DNA</u>

1. Inoculate 100 mL LB (plus antibiotic) in a 1-L flask with 50 to 100 µL of an overnight culture of transformed bacteria and grow the culture to saturation by shaking overnight at 37°C.
2. Harvest the cells by centrifugation for 5 minutes at 4000× g (e.g., 5000 rpm in a GSA rotor).
3. Remove the supernatant, drain well, and resuspend the pelleted cells in 3 mL Solution I. Vortex mix or pipet up and down to be sure that the cells are evenly and completely resuspended.
4. Add 1 mL 8 mg/mL lysozyme solution, mix gently, and incubate on ice for 5 minutes.
5. Add 8 mL Solution II, mix gently, and incubate on ice for 5 minutes. The solution will become quite viscous as the cells are lysed.
6. Add 6 mL Solution III, mix gently, and incubate on ice for 15 minutes. A large, gloppy precipitate will form.
7. Pellet cell debris and insoluble material by centrifugation at 12 000× g (e.g., 8500 rpm in a GSA rotor) for 20 minutes at 4°C.
8. Collect the supernatant by pouring it into a clean centrifuge tube, through several layers of cheesecloth. This will remove stray pieces of precipitated material that may be floating free from the pellet.
9. Add 2 to 3 volumes of EtOH and incubate at -20°C for at least 30 minutes.
10. Pellet the precipitated DNA by centrifugation at 12 000× g (e.g., 8500 rpm in a GSA rotor) for 20 minutes at 4°C.
11. Pour off and discard the supernatant. Wipe excess EtOH from the sides of the centrifuge bottle and allow the pellet to air dry. Alterna-

tively, dry the pellet briefly under a vacuum.

12. Dissolve the pelleted DNA in 5 mL TE (50:50).

<u>Band the DNA in CsCl/EtBr Gradient:</u>

1. Transfer exactly 5 mL of the DNA solution to a centrifuge tube (e.g., 30-mL Corex) and add 5.5 g CsCl and 0.25 mL 10 mg/mL EtBr. Mix well to dissolve the CsCl completely and then let the solution sit at room temperature for 15 minutes.

2. Centrifuge the DNA/CsCl at $12\,000\times g$ (e.g., 10 000 rpm in an SS-34 rotor) for 10 minutes at room temperature to remove precipitated material.

3. Load the supernatant into a sealable ultracentrifuge tube. Centrifuge for at least 6 hours (or overnight if convenient) at 55 000 rpm at 20°C in a VTi65.1 rotor (Beckman Instruments, Fullerton, CA, USA). Use a slow deceleration program to minimize band disruption as the rotor slows and the gradient reorients.

Notes: (*i*) Don't be careless with volume and weight measurements in assembling the DNA/CsCl/EtBr solution before banding because the separation is density based. The plasmid DNA should band near the center of the ultracentrifuge tube. If the band is too high in the tube, it indicates that the starting solution was too dense; likewise, if it is too low, the solution was not dense enough. (*ii*) The plasmid DNA will band several millimeters below the chromosomal DNA, which is not always visible, depending on how much was eliminated during preparation of the cleared lysate. RNA will be at the bottom of the tube, and protein will complex with EtBr and form a skin at the top of the gradient. Since the gradient is formed vertically in vertical rotors, this skin sometimes remains stuck along the side of the tube when the gradient reorients as the rotor stops spinning. Care must be taken to avoid the protein skin when collecting the plasmid band. (*iii*) The specific conditions for centrifugation will depend on the equipment available. The DNA can be banded in different volumes or at higher speeds for less time. Refer to the manufacturer's instructions for particular centrifuges and rotors. (*iv*) Ethidium bromide is a mutagen and should be handled with appropriate personal protection (e.g., gloves, laboratory coat) and waste should be disposed of according to institutional guidelines.

<u>Recover DNA from the CsCl/EtBr Gradient</u>

1. Insert an 18-gauge needle into the top of the sealed ultracentrifuge tube to create an air vent. Alternatively, if using OptiSeal tubes (Beckman In-

Table 1. Summary of Steps for CsCl/EtBr Gradient Purification of Plasmid DNA

<u>Day 1</u>

1. Inoculate 100 mL LB (plus antibiotic) with 50–100 µL overnight culture of transformed bacteria.
2. Grow to saturation by shaking overnight at 37°C.

<u>Day 2</u>

Precipitate plasmid DNA from cleared lysate

1. Harvest cells by centrifugation for 5 minutes at 4000× g.
2. Resuspend pelleted cells in 3 mL Solution I.
3. Add 1 mL 8 mg/mL lysozyme solution; incubate on ice for 5 minutes.
4. Add 8 mL Solution II; incubate on ice for 5 minutes.
5. Add 6 mL Solution III; incubate on ice for 15 minutes.
6. Centrifuge at 12 000× g for 20 minutes at 4°C.
7. Collect supernatant by pouring through cheesecloth. Add 2–3 volumes EtOH and incubate at -20°C for 30 minutes.
8. Pellet precipitated DNA by centrifugation at 12 000× g for 20 minutes at 4°C.
9. Resuspend the DNA in 5 mL TE (50:50)

Band the DNA in CsCl/EtBr

1. Add 5.5 g CsCl and 0.25 mL 10 mg/mL EtBr to exactly 5 mL of DNA in TE (50:50). Dissolve thoroughly and let stand at room temperature for 15 minutes.
2. Centrifuge 12 000× g for 10 minutes at room temperature.
3. Load supernatant into sealable ultracentrifuge tube.
4. Centrifuge to band DNA (e.g., 55 000 rpm at 20°C overnight in a VTi65.1 rotor).

<u>Day 3</u>

Recover the banded DNA and reband (optional)

1. Pull the DNA band from the ultracentrifuge tube using a syringe.
2. Combine the DNA with CsCl/EtBr/TE (50:50) to fill a new ultracentrifuge tube.
3. Centrifuge to band the DNA (e.g., 55 000 rpm at 20°C for 6 hours in a VTi65.1 rotor).
4. Pull the DNA band from the tube, as above.

Remove EtBr and CsCl from the DNA

1. Add an equal volume of isoamyl alcohol to the DNA. Vortex mix for 1 minute.
2. Centrifuge briefly to separate phases, collect the aqueous phase, and repeat for a total of 3–4 extractions.

Table 1. (Continued)

> 3. After the final extraction, add an equal volume of water to the aqueous phase. Add 2 volumes (relative to the new total volume) of EtOH and incubate for 1 hour in the refrigerator.
> 4. Pellet DNA by centrifugation at $12\,000\times g$ for 20 minutes. Pour off supernatant, drain well, and dry briefly.
> 5. Resuspend the DNA TE^{-4}, transfer to a microfuge tube, and reprecipitate by adding 1/10 volume 3 mol/L sodium-acetate and 2–3 volumes of EtOH. Incubate on dry ice for 5 minutes.
> 6. Collect the precipitated DNA by centrifugation for 10 minutes. Aspirate the supernant and briefly dry the pellet. Resuspend the DNA in TE^{-4}.

struments), remove the plug from the top of the tube. It may be helpful to illuminate the tube with long wave UV light to visualize the banded chromosomal and plasmid DNA.

2. Using a second 18-gauge needle attached to a 5-mL syringe, pierce the tube approximately 5 mm below the banded plasmid DNA and place the tip of the needle, with the bevel facing up, at the bottom of the band.

3. Slowly draw the banded DNA into the syringe, being careful to avoid turbulence that could cause mixing of the plasmid and chromosomal DNA bands. The band can usually be recovered in a volume of 1 mL or less.

4. Withdraw the syringe, remembering that the tube contents will now drip from the hole left by the syringe. Place protective sheeting and a receptacle below the tube, or quickly cover the hole with a gloved finger and move the entire tube to a waste receptacle.

Reband the DNA in a Second CsCl/EtBr Gradient

This step is optional, but recommended, since it is useful in removing residual contaminants such as RNA, chromsomal DNA, proteins, and endotoxins, that may be present to varying degrees after the first banding.

1. Prepare a blank solution of CsCl/EtBr by combining 5 mL TE (50:50), 5.5 g CsCl, 0.25 mL 10 mg/mL EtBr. Multiply these amounts as necessary to make an appropriate total volume.

2. Add DNA collected from the first gradient to a new sealable ultracentrifuge tube and fill the tube with blank CsCl/EtBr solution.

3. Seal the tube, centrifuge, and collect the banded DNA as above.

Remove EtBr and CsCl from the DNA

1. Transfer the DNA/CsCl solution to a clear centrifuge tube (e.g., glass Corex) and add an equal volume of isoamyl alcohol. Vortex mix for approximately 1 minute and then separate the phases by brief (1 min) centrifugation at about 1000× g in a tabletop centrifuge with a swinging bucket rotor. The phases separate easily without centrifugation, but the spin helps to clean the sides of the tube and ultimately allows a more thorough removal of EtBr.

2. Remove the aqueous phase (that contains the DNA) and repeat for a total of 3 to 4 extractions with isoamyl alcohol, until the aqueous phase is completely colorless.

3. Save the aqueous phase of the final extraction and add an equal volume of water. Then add two volumes (relative to the new total volume) of EtOH and incubate for up to 1 hour in the refrigerator.

4. Pellet the precipitated DNA by centrifugation at 12 000× g for 20 minutes. Drain the supernatant well and air-dry briefly.

5. Dissolve the DNA pellet in a small volume (e.g., 250 µL) of TE^{-4}, transfer to a microcentrifuge tube, then reprecipitate once by adding 1/10 volume 3 M sodium-acetate and 2 to 3 volumes EtOH. Incubate on dry ice for 5 minutes.

6. Pellet the DNA by centrifugation in a microcentrifuge for 5 to 10 minutes. Rinse the pellet with 70% EtOH and allow the pellet to air-dry. Resuspend the DNA in TE^{-4} at a convenient concentration, usually equal to or greater than 1 mg/mL.

7. DNA can be stored refrigerated for short periods (days). For longer storage, DNA should be kept frozen at -20° or -80°C, although it is important to avoid repeated freeze-thaw cycles.

Notes: (*i*) The organic extraction should be performed in a fume hood, and waste handled and disposed of according to institutional guidelines. Alternatively, isopropanol that has been pre-equilibrated with a saturated solution of NaCl may be used as the organic phase for extractions. (*ii*) To minimize precipitation of the CsCl at step 3, do not add more than 2 volumes of EtOH, incubate at temperatures colder than the refrigerator, or for periods of time longer than about 2 hours. If the yield of DNA is high, a precipitate will form quickly and no cold incubation is required at this step.

Purification of Plasmid DNA by Anion Exchange Chromatography

The following procedure is based on instructions for the use of columns manufactured by Qiagen (Valencia, CA, USA). Several manufacturers supply similar kits with detailed procedures for their use. In each case, users should follow the directions supplied with the kit and pay attention to special notes and options (such as removal of endotoxins) at some steps.

Prepare Cleared Lysate and Precipitate Plasmid DNA

1. Inoculate 100 mL LB (plus antibiotic) with 50 to 100 μL of an overnight culture of transformed bacteria and grow the culture to saturation by shaking overnight at 37°C.
2. Harvest the cells by centrifugation for 5 minutes at 4000× g (e.g., 5000 rpm in a GSA rotor).
3. Remove the supernatant, drain well, and resuspend the pelleted cells in 10 mL Buffer P1. Vortex mix or pipet up and down to be sure that the cells are evenly resuspended.
4. Add 10 mL Buffer P2, mix gently, and incubate at room temperature for 5 minutes.
5. Add 10 mL cold Buffer P3, mix gently, and incubate on ice for 20 minutes.
6. Centrifuge at 20 000× g (e.g., 13 000 rpm in an SS-34 rotor) for 30 minutes at 4°C to pellet the cell debris, chromosomal DNA, RNA, and proteins that were precipitated by the SDS and potassium-acetate.
7. Collect the supernatant. If some precipitated material remains floating in the supernatant, it can be removed by pouring through several layers of cheesecloth or by a second centrifugation at 20 000× g for 15 to 30 minutes at 4°C.

Notes: (*i*) The above procedure is based on a high copy number plasmid and a yield of 500 μg DNA. The volume of culture needed will depend on the plasmid copy number and the capacity of the anion exchange column that will be used. Since yield can be reduced by overloading, the column size must be matched with an appropriate culture volume. (*ii*) The lysed cells should be mixed gently after addition of Buffers P2 and P3 to avoid shearing the chromosomal DNA. Small fragments of DNA will remain soluble in the lysate and copurify with the plasmid during anion exchange chomotography.

Anion Exchange Chromatography

1. Equilibrate a Qiagen-tip 500 with 10 mL Buffer QBT. Allow the buffer to flow through by gravity.
2. Apply the cleared lysate (from step 7 above) to the column and allow it

to flow into the resin by gravity.

3. Wash the column with 30 mL Buffer QC, allowing it to flow through the resin by gravity. Repeat the wash with an additional 30 mL Buffer QC.

4. Use 15 mL Buffer QF to elute the DNA, collecting the eluate (by gravity flow) in a 50-mL centrifuge tube.

5. Add 0.7 volumes (10.5 mL) isopropanol to the eluate to precipitate the DNA.

6. Collect the precipitated DNA by centrifugation at 20 000× g for 20 minutes. Drain the pellet well.

7. Wash the pellet once by adding 70% ethanol and centrifuging at 20 000× g for 10 minutes.

8. Air-dry (or briefly dry under vacuum) and resuspend the pelleted DNA in TE^{-4} at a convenient concentration, usually equal to or greater than 1 mg/mL.

Media and Solutions

<u>LB medium (1 L)</u>

10 g tryptone
5 g yeast extract
10 g NaCl
1 mL 1N NaOH

Autoclave for 30 minutes. Antibiotics are added after autoclaving. The choice of antibiotic depends on the resistance gene present on the plasmid, but some commonly used antibiotics can be added with final concentrations as follows:

ampicillin: 100 µg/mL
kanamycin: 40 µg/mL
tetracycline: 15 µg/mL

For plates, add 16 g agar per liter of medium.

<u>CaCl$_2$/PIPES solution</u>

50 mM CaCl$_2$
5 mM PIPES, pH 6.8

<u>SOB (1 L)</u>

20 g tryptone

5 g yeast extract
0.6 g NaCl
0.6 g KCl
Autoclave ingredients in 990 mL water. Cool, then add 5 mL 1 M $MgCl_2$ and 5 mL 1 M $MgSO_4$. Filter-sterilize the Mg solutions prior to adding.

SOC

Add 1 mL 2 M glucose (filter-sterilized) to 100 mL SOB.

FSB/glycerol

10 mM K-4-morpholineethanesulfonic acid (MES), pH 6.3
100 mM KCl
45 mM $MnCl_2$
10 mM $CaCl_2 \cdot 2H_2O$
3 mM hexamine $CoCl_3$
10% glycerol
Prepare a stock of 0.5 M MES (free acid) and adjust the pH to 6.3 with KOH. Filter-sterilize the FSB solution.

Solution I

50 mM glucose
10 mM EDTA
25 mM Tris-HCl, pH 8.0

Lysozyme solution

8 mg/mL lysozyme in Solution I

Solution II = Buffer P2

0.2 N NaOH
1% SDS

Solution III = Buffer P3

3 M potassium-acetate, pH 5.5
Adjust pH with glacial acetic acid.

TE (50:50)

50 mM EDTA
50 mM Tris-HCl, pH 8.0

TE^{-4}

10 mM Tris-HCl, pH 8.0
0.1 mM EDTA
This is simply TE (10 mM Tris-HCl, 1 mM EDTA, pH 8.0) with 1/10 as much EDTA. TE will substitute for TE^{-4} in all cases.

Buffer P1

50 mM Tris-HCl, pH 8.0
10 mM EDTA
100 µg/mL RNaseA
The solution is stable for several months if stored at 4°C after the addition of RNaseA.

Buffer QBT

750 mM NaCl
50 mM 4-morpholinepropanesulfonic acid (MOPS), pH 7.0
15% isopropanol
0.15% Triton® X-100

Buffer QC

1.0 M NaCl
50 mM MOPS, pH 7.0
15% isopropanol

Buffer QF

1.25 M NaCl
50 mM Tris-HCl, pH 8.5
15% isopropanol

REFERENCES

1. **Birnboim, H.C.** 1983. A rapid alkaline extraction method for the isolation of plasmid DNA. Methods Enzymol. *100*:243-255.
2. **Birnboim, H.C., and J. Doly.** 1979. A rapid alkaline extraction procedure for screening recombinant plasmid DNA. Nucleic Acids Res. *7*:1513-1523.
3. **Clewell, D.B., and D.R. Helinski.** 1969. Supercoiled circular DNA-protein complex in *Escherichia coli*: Purification and induced conversion to an open circular DNA form. Proc. Natl. Acad. Sci. USA *62*:1159-1166.
4. **Cohen, S.N., A.C.Y. Chang, and L. Hsu.** 1972. Nonchromosomal antibiotic resistance in bacteria: genetic transformation of *Escherichia coli* by R-factor DNA. Proc. Natl. Acad. Sci. USA *69*:2110-2114.
5. **Dagert, M. and S.D. Ehrlich.** 1979. Prolonged incubation in calcium chloride improves the competence of *Escherichia coli* cells. Gene *6*:23-28.
6. **Dower, W. J., J.F. Miller, and C.W. Ragsdale.** 1988. High efficiency transformation of *E. coli* by high voltage electroporation. Nucleic Acids Res. *16*:6127-6145.
7. **Hanahan, D.** 1983. Studies on transformation of *Escherichia coli* with plasmids. J. Mol. Biol. *166*:557-580.
8. **Holmes, D.S. and M. Quigley.** 1981. A rapid boiling method for the preparation of bacterial plasmids. Anal. Biochem. *114*:193-197.
9. **Inoue, H., H. Nojima, and H. Okayama.** 1990. High efficiency transformation of *Escherichia coli* with plasmids. Gene *96*:23-28.
10. **Kushner, S.R.** 1978. An improved method for transformation of *Escherichia coli* with ColE1-derived plasmids, p. 17-23. *In* H.W. Boyer and S. Nicosia (Eds.), Genetic Engineering: Proceedings of the International Symposium on Genetic Engineering. Elsevier, Amsterdam.
11. **Mandel, M. and A. Higa.** 1970. Calcium-dependent bacteriophage DNA infection. J. Mol. Biol. *53*:159-162.
12. **Oishi, M., and S.D. Cosloy.** 1972. The genetic and biochemical basis of the transformability of *Escherichia coli* K12. Biochem. Biophys. Res. Commun. *49*:1568.
13. **Radloff, R., W. Bauer, and J. Vinograd.** 1967. A dye-buoyant-density method for the detection and isolation of closed circular duplex DNA: the closed circular DNA in HeLa cells. Proc. Natl. Acad. Sci. USA *57*:1514-1521.
14. **Sambrook, J., E.F. Fritsch, and T. Maniatis.** 1989. Molecular Cloning: a Laboratory Manual, 2nd ed. CSH Laboratory Press, Cold Spring Harbor. NY.
15. **Taketo, A.** 1988. DNA transfection of *Escherichia coli* by electroporation. Biochim. Biophys. Acta *949*:318-324.
16. **Tang, X., Y. Nakata, H.-O. Li, M. Zhang, H. Gao, A. Fujita, O. Sakatsume, T. Ohta, and K. Yokoyama.** 1994. The optimization of preparations of competent cells for transformation of *E. coli*. Nucleic Acids Res. *22*:2857-2858.

Yeast Transformation

Robin A. Woods[1] and R. Daniel Gietz[2]
[1]Department of Biology, University of Winnipeg, Winnipeg, Manitoba; [2]Department of Human Genetics, University of Manitoba, Winnipeg, Manitoba, Canada

INTRODUCTION

The transformation of the yeast *Saccharomyces cerevisiae* with exogenous DNA involves three sequential processes: *(i)* pretreatment of the cells before transformation to make them competent (able to take up DNA); *(ii)* incubation with transforming DNA in a transformation mixture; and finally *(iii)* plating onto medium that will select for the desired transformants. There are three methods by which transformation can be accomplished: *(i)* spheroplast transformation, *(ii)* electroporation, and *(iii)* the lithium acetate/single-stranded carrier DNA polyethylene glycol (LiAc/SS carrier DNA/PEG) procedure. The last of these is the most commonly employed by yeast molecular biologists and is the focus of this chapter.

The transformation of yeast was first reported in 1978 (2,13). It involved the conversion of cells to spheroplasts by enzymatic digestion of the cell wall, stabilization of the spheroplasts with sorbitol, and then transformation by plasmid DNA in the presence of polyethylene glycol (PEG 4000). With careful attention to the treatment of spheroplasts and the concentrations of plasmid and carrier DNA, this procedure yields up to 3.3×10^5 transformants/µg double-stranded DNA plasmid/10^8 cells (4). Spheroplast transformation has become the method of choice for the manipulation of yeast artificial chromosomes in yeast and has been optimized for this purpose (21). The main drawbacks of this procedure are that: *(i)* cell wall digestion must be carefully monitored, *(ii)* the fragile spheroplasts must be stabilized in 1 M sorbitol, and *(iii)* after transformation, the spheroplasts must be suspended in regeneration agar plus sorbitol and poured onto dishes of selective

Gene Transfer Methods
Edited by Pamela A. Norton and Laura F. Steel
©2000 Eaton Publishing, Natick, MA

medium. This means that all transformants must be isolated manually or sampled with a custom-built multi-pin replicating device. Alternatively, the transformed spheroplasts can regenerate in a thin layer of calcium alginate; this allows the colonies to be replica-plated by standard techniques (23).

Transformation by electroporation is the least efficient method of the three. However, it involves the least experimental time, and one procedure requires only 10 minutes to transform cells grown overnight on agar plates (12). Most of the protocols reported involve growing cells to an optical density of 1.3 to 1.5 at 600 nm; corresponding to about 1×10^8 cells/mL or mid-logarithmic phase of growth (1,15,20). The subsequent treatment of the cells is variable. The cells may be washed in water and 1 M sorbitol, electroporated in 0.5 M sorbitol, resuspended in 1 M sorbitol, and plated onto selective medium (1) or washed and electroporated in a buffer (0.3 mM NaH_2PO_4, 0.2 mM KH_2PO_4, 10% glycerol) containing carrier DNA (20). One protocol that yields 2.5×10^5 transformants/μg plasmid DNA/10^8 cells borrows from the LiAc/SS carrier DNA/PEG protocol in that the cells are incubated at 42°C for 15 minutes with plasmid, SS carrier DNA, and PEG 4000 and then electroporated (15).

Transformation of yeast cells after exposure to alkali cations was reported in 1983 (14) and soon became the method of choice in many laboratories. The treated cells were incubated with plasmid DNA and PEG 4000 at 30°C and then plated onto selective medium after a 5-minute heat shock at 42°C. The yield of transformants, 400/μg plasmid DNA, was low relative to contemporary spheroplast transformation (22), but the technique was simpler, faster, and more reproducible. Li$^+$, as lithium acetate (LiAc), is the most effective alkali cation (14). The addition of sonicated double-stranded (DS) carrier DNA (7,17) to the transformation mixture increased efficiency to 2×10^4 transformants/μg plasmid/10^8 cells. Efficiencies within the same range as the spheroplast procedure were obtained by the use of SS rather than DS carrier DNA (18). The LiAc/SS carrier DNA/PEG protocol (18) involved dilution of an overnight culture into fresh medium and regrowth for two divisions. The cells were centrifuged and resuspended in Tris-EDTA/LiAc buffer at 30°C for 60 minutes. Plasmid and SS carrier DNA were added, and incubation continued for another 30 minutes. PEG 4000 (50% wt/vol) was added to a final concentration of 33%. After a further 30 minutes at 30°C, the suspension of cells in this transformation mixture was incubated at 42°C for 15 minutes, centrifuged, resuspended in Tris-EDTA (TE) buffer, and plated onto selective medium. This procedure gave yields up to 5×10^4 transformants/μg plasmid DNA/10^8 cells (18).

We have improved and streamlined the protocol by reducing exposure to LiAc before transformation (8), omitting TE from the protocol entirely (9), and optimizing the number of cells and concentrations of carrier DNA and plasmid in the transformation mixture (10). Cultures for transformation can be incubated overnight on agar plates and then grown in broth medium for two to four divisions before transformation (19).

All three methods of transformation involve modification of the cell wall or cell membrane to facilitate the uptake of exogenous DNA. Conversion of cells to spheroplasts involves removal of the cell wall. Electroporation is thought to result in a temporary breakdown of the cell membrane and the formation of pores that allow macromolecules to enter or leave the cell (16). Treatment of intact yeast cells with Li^+ and sodium dodecyl sulfate (SDS) results in leakage of nucleic acids, suggesting that the membrane becomes permeable (3). Treatment with PEG is an integral component of the spheroplast and alkali cation methods and is required for the most effective electroporation protocol. We have shown that PEG deposits carrier and plasmid DNA onto cells, and that incubation of the suspension of cells in transformation mixture at 42°C for at least 20 minutes is essential for optimal recovery of transformants (10).

Transformation can be used to study yeast molecular biology as well as the expression of foreign genes in yeast. Generally, a researcher is concerned with the selection of a single type of transformant from a yeast gene library or with the transformation of a specific plasmid into one or more yeast strains. A yeast genomic library contains about 6000 clones, and one needs to screen approximately 20 000 clones to ensure complete coverage. The rapid LiAc/SS carrier DNA/PEG transformation protocol described below easily generates 40 000 transformants/μg plasmid DNA/10^8 cells, sufficient for such experiments.

The transformation of complex gene libraries into yeast, as is required for systems such as for the one- and two-hybrid screens that detect protein–DNA and protein–protein interactions (11), necessitates the highest transformation yields. The high-efficiency LiAc/SS carrier DNA/PEG transformation protocol yields 2 to 5×10^6 transformants/μg plasmid DNA/10^8 cells and can be scaled up to give the numbers required to screen complex libraries.

PROTOCOLS

LiAc/SS Carrier DNA/PEG Transformation

The rapid, high-efficiency and two-hybrid screen LiAc/SS carrier DNA/PEG transformation protocols all make use of a basic transformation mixture. The recipe below is sufficient for the transformation of 1×10^8 cells.

Component	Volume (µL)
PEG 3350 (50% wt/vol) (Sigma [St. Louis, MO, USA] revised the molecular weights of their PEG preparations, and PEG 4000 was renamed PEG 3350)	240
LiAc 1.0 M	36
SS carrier DNA (2.0 mg/mL)	50
Plasmid DNA (100 ng) plus water (distilled/deionized)	34
Total volume (excluding cells)	360

If a number of transformations are planned, this transformation mixture can be prepared in bulk and kept in ice water until required. The highest transformation efficiencies (transformants/µg plasmid DNA/10^8 cells) are obtained with 100 ng plasmid DNA, but the number of transformants will be greater if more plasmid is used.

Rapid LiAc/SS Carrier DNA/PEG Transformation Protocol

This protocol can be used when large numbers of transformants are not required. Yields for most strains should be of the order of 4×10^4 transformants/µg plasmid/10^8 cells or about 4000 per individual transformation.

1. Inoculate the yeast strain to be transformed onto a plate of YPAD agar (see section entitled Media for components of this medium) in a patch approximately 2 cm^2 and incubate overnight at 30°C.

2. The next day, place a tube of carrier DNA (2.0 mg/mL) in a boiling water bath for 5 minutes and then chill in ice water. This step is given first, because if you forget to boil the carrier DNA the transformation will be delayed for 10 to 15 minutes.

3. Scrape about 20 µL (2×10^8) of cells from the overnight culture with a sterile toothpick or inoculating loop and resuspend them in 1 mL sterile water in a 1.5-mL microcentrifuge tube.

4. Pipet samples of 500 µL (1×10^8 cells) for each transformation into 1.5-mL microcentrifuge tubes. Centrifuge at top speed for 15 seconds and remove the supernatant.

5. Vortex mix the SS carrier DNA and make up enough transformation mixture for the number of planned transformations plus one extra. Vortex mix the bulk vigorously and keep it in ice water.

6. Pipet 360 µL of transformation mixture onto each cell pellet and vortex mix vigorously to resuspend the cells.

7. Incubate at 42°C for 60 minutes.

8. Centrifuge and remove the supernatant.

Table 1. The Rapid Transformation Protocol

1. Grow strain overnight on YPAD agar plate.
2. Boil carrier DNA for 5 minutes and chill.
3. Suspend approximately 20 µL of cells in 1.0 mL water in microcentrifuge tube, centrifuge, and discard water.
4. Resuspend in 1.0 mL water and pipet half of the suspension into another tube.
5. Centrifuge both tubes and discard the supernatants.
6. Add transformation mixture (PEG 3350 240 µL, LiAc 36 µL, carrier DNA 50 µL, plasmid 100 ng, water to 360 µL) to the cell pellets and vortex mix.
7. Incubate at 42°C: 60 minutes for 10 000 transformants/tube, 120 minutes for 40 000 transformants/tube, and 180 minutes for 50 000 transformants/tube.
8. Centrifuge and discard the transformation mixture.
9. Resuspend the cells in 1.0 mL water and plate samples onto SC minus

9. Pipet 1.0 mL water onto the cell pellet and resuspend the cells by pipetting up and down. Vortex mix vigorously to ensure complete dispersal of the cells.

10. Plate samples of 10 and 100 µL onto plates of SC minus selection agar (see section entitled Media for components of medium) and incubate for 3 days. The 10-µL sample should be pipetted into a 100-µL puddle of sterile water on the agar surface and then spread.

The steps of this protocol are summarized in Table 1. The duration of incubation at 42°C and the age of the culture affect the yield of transformants. Table 2 shows the results of incubating cells from 1-, 2-, and 4-day cultures at 42°C for up to 240 minutes. The highest yields are obtained if the transformation reaction is incubated at 42°C for 180 minutes. Similar transformation efficiencies are observed with cells grown overnight in 5 or 10 mL of YPAD or 2× YPAD broth to a titer of approximately 2×10^8 cells/mL.

High-Efficiency LiAc/SS Carrier DNA/PEG Transformation Protocol

This protocol should give 2 to 5×10^6 transformants/µg plasmid DNA/10^8 cells with strains that transform well.

1. Inoculate the yeast strain onto a plate of YPAD agar in a patch about 2 cm^2 or into 5 mL of 2× YPAD in a 15-mL culture tube. Incubate the plate, a bottle of double strength YPAD broth (2× YPAD) and a sterile

Table 2. The Rapid Transformation Protocol: Effects of Culture Age and Duration of Incubation at 42°C on Transformation Efficiency

Minutes at 42°C	Transformants/µg Plasmid DNA/10^8 Cells		
	1-Day Culture	2-Day Culture	4-Day Culture
0	0	0	0
60	211 180	48 260	19 490
120	785 210	136 100	108 060
180	1 248 800	308 880	125 780
240	889 340	279 920	158 900
300	560 060	254 830	124 010

Strain DY2389 (*α ade2 can1 his3 leu2 lys2 trp1 ura3*), kindly provided by Dr. B. Stillman, Health Science Center, University of Utah) was grown on YPAD plates and transformed with plasmid YCplac33 (7) using the rapid transformation protocol. Transformants were selected on SC medium minus uracil.

250-mL flask overnight at 30°C. The culture tube should be incubated on a rotary shaker at approximately 200 rpm.

2. Scrape about 100 µL of cells from the YPAD plate with a sterile toothpick or inoculating loop and resuspend them in 1 mL sterile water in a 1.5-mL microcentrifuge tube. The broth culture should have grown to a titer of about 2×10^8 cells/mL.

3. Determine the titer of the suspension or culture by counting the cells in a hemocytometer or measuring the OD at 545 or 600 nm of a 10 µL to 1.0 mL dilution. For most yeast strains, 1.0×10^6 cells/mL gives an OD_{545} of about 1.0. To calculate accurate transformation efficiencies, determine the relationship between the hemocytometer count or OD and the number of viable cells by plating appropriate dilutions onto YPAD agar plates.

4. Pipet 50 mL of the warm 2× YPAD broth into the sterile flask and add 2.5×10^8 cells to give a final titer of 5×10^6 cells/mL.

5. Incubate the flask on a rotary shaker at 30°C and 200 rpm until the cells have gone through two divisions. This will take 4 to 5 hours. The cell titer should be about 2×10^7 cells/mL and the total cell number about 1×10^9. This number of cells is sufficient for 10 standard transformations.

6. Place a tube of carrier DNA (2.0 mg/mL) in a boiling water bath for 5 minutes and then chill in ice water.

7. Harvest the culture in a 50-mL centrifuge tube. Centrifugation at 3000 to 4000 rpm (approximately 1500× g) for 5 to 10 minutes in a bench centrifuge (e.g., IEC Centra MP4 centrifuge; IEC, Needham Heights, MA, USA) is sufficient.

8. Resuspend the cells in 25 mL sterile water, centrifuge again, and resuspend in 1 mL sterile water.

9. Transfer to a 1.5-mL microcentrifuge tube, pellet the cells, and resuspend them to a final volume of 1.0 mL.

10. Pipet 100-µL samples of 1×10^8 cells for each transformation into 1.5-mL microcentrifuge tubes. Centrifuge at top speed for 15 seconds and remove the supernatant.

11. Vortex mix the SS carrier DNA and make up enough transformation mixture for the number of planned transformations plus one extra. Vortex mix the transformation mixture vigorously and keep it in ice water.

12. Pipet 360 µL of transformation mixture onto each pellet and vortex mix vigorously to resuspend the cells.

13. Incubate at 42°C for 40 minutes.

14. Centrifuge and remove the supernatant.

15. Pipet 1.0 mL water onto the cell pellet and resuspend the cells by pipetting up and down. Vortex mix vigorously to ensure complete dispersal of the cells.

16. Pipet samples of 1 and 10 µL into 100-µL puddles of sterile water on plates of SC minus selection agar; spread the plated cells with a sterilized glass spreader and incubate for 3 days.

The steps of this protocol are summarized in Table 3. The optimal duration of incubation at 42°C for the regrown cells used in this protocol is 40 minutes (see Table 4), much shorter than for cells grown to stationary phase.

Two-Hybrid Screen Transformation Protocol

Screens to test for protein–protein or protein–DNA interactions usually involve the transformation of two plasmids into one yeast strain. In the two-hybrid and similar screens (5,6,11), the first plasmid carries the DNA binding domain (*GAL4_BD*) of the yeast *GAL4* gene. This domain is fused, in frame, to the coding region of a specific gene, typically referred to by yeast workers as *YFG* (your favorite gene). This plasmid can be transformed into a suitable yeast strain by the rapid transformation protocol. The second transformation involves plasmids that carry the *GAL4* transcription-activating do-

Table 3. The High-Efficiency Transformation Protocol

1. Inoculate the yeast strain in a 2 cm^2 patch on a YPAD agar plate or into 5 mL YPAD in a culture tube and incubate overnight at 30°C. The tube culture should be in a shaking incubator.

2. Warm 2× YPAD broth and culture flask overnight.

3. Suspend cells from the plate in 1.0 mL water and determine the cell titer; determine the titer of the 5-mL culture.

4. Inoculate 50 mL warm 2× YPAD broth with 2.5×10^8 cells.

5. Incubate on shaker for at least 4 hours until the cell titer is at least 2×10^7/mL.

6. Boil carrier DNA for 5 minutes and chill.

7. Harvest cells, wash them in 25 mL water, and resuspend them in 1 mL water.

8. Transfer the suspension to a 1.5-mL microcentrifuge tube, pellet the cells, and resuspend them to a final volume of 1.0 mL.

9. Dispense 100-µL samples into microcentrifuge tubes, centrifuge, and discard the supernatant.

10. Make bulk transformation mixture (PEG 3350 240 µL, LiAc 36 µL, carrier DNA 50 µL, plasmid 100 ng, water to 360 µL) for each transformation plus one extra.

11. Add 360 µL transformation mixture to each pellet and vortex mix to resuspend cells.

12. Incubate at 42°C for 40 minutes.

13. Centrifuge and remove the transformation mixture.

14. Resuspend cells in 1.0 mL water and plate samples onto SC minus selection agar.

main ($GAL4_{AD}$) fused to cDNA sequences derived from cellular mRNA—a $GAL4_{AD}$/cDNA plasmid library. Interaction between the protein encoded by YFG and the translation product of a cDNA sequence positions the $GAL4_{BD}$ and $GAL4_{AD}$ domains such that they activate a reporter gene and cause the expression of a specific phenotype. Details of the construction and testing of these fusion plasmids and the genotypes of suitable yeast strains can be found in Gietz et al. (11).

Productive protein–protein interactions are rare and require the screening of millions of cells carrying the $GAL4_{BD}$/YFG fusion plasmid after transformation with the $GAL4_{AD}$/cDNA plasmid library. This second transformation is not as efficient as transformation with a single plasmid, and yields are much reduced. The high-efficiency transformation protocol must be scaled

Table 4. The High-Efficiency Transformation Protocol: Effect of Duration of Incubation at 42°C on Transformation Efficiency

Minutes at 42°C	Transformants/µg Plasmid DNA/10^8 Cells
0	8.4×10^3
10	2.2×10^5
20	1.8×10^6
30	3.8×10^6
40	5.8×10^6
50	3.5×10^6
60	3.4×10^6

Strain DY2389 was grown overnight in 10 mL 2× YPAD to 1.96×10^8 cells/mL, diluted to 5×10^6/mL in 50 mL warm 2× YPAD, and regrown for 4.5 hours to 1.8×10^7/mL. Cells were transformed with plasmid YCplac33 using the high-efficiency transformation protocol. Transformants were selected on SC medium minus uracil.

up 30×, 60×, or even 120×, to ensure the recovery of sufficient numbers of transformants. Note that the strain carrying the *GAL4*$_{BD}$/*YFG* fusion plasmid must be maintained on selective medium to ensure retention of the plasmid; however, it can be grown in 2× YPAD broth, without significant loss of the plasmid, for the two divisions prior to transformation with the *GAL4*$_{AD}$/cDNA plasmid library.

The sequence of operations involved in a two-hybrid screen is as follows:

1. Transform the *GAL4*$_{BD}$/*YFG* fusion plasmid into a suitable yeast strain using the rapid transformation protocol and select primary transformants on plates of SC minus selection agar.

2. Inoculate a primary transformant into SC minus selection liquid medium and incubate at 30°C overnight on a rotary shaker at approximately 200 rpm. Warm a bottle of 2× YPAD broth and a culture flask overnight at 30°C.

3. Use the high-efficiency transformation protocol to determine the yields of secondary transformants with a range of concentrations or volumes of the *GAL4*$_{AD}$/cDNA plasmid library. If the concentration of the plasmid library is known, we suggest a range from 100 ng to 10 µg; if not, then try a range of volumes from 1 to 10 µL. Determine how much plasmid per unit transformation will maximize the number of transformants, and then scale up the protocol accordingly. In the steps below, we give the volumes of medium and reagents for 30× and 60× scale up experiments.

4. Inoculate a primary transformant into SC minus selection liquid medium and incubate at 30°C overnight on a rotary shaker at approximately 200 rpm (30× — 50 mL medium/250-mL flask; 60× — 100 mL medium/500-mL flask). Warm 2× YPAD broth (30× — 200 mL; 60× — 400 mL) and culture flasks (30× — 1 L; 60× — 2 × 1 L) overnight at 30°C.

5. Determine the titre of the overnight culture and calculate the volume that will contain 7.5×10^8 cells (30×) or 1.5×10^9 cells (60×). Pour or pipet this volume into one or more 50-mL centrifuge tubes and pellet the cells by centrifugation at 1500× g for 5 to 10 minutes. Resuspend the cells in warm 2× YPAD broth and transfer to the culture flask(s). Add 2× YPAD broth to bring the cell titer to 5×10^6/mL (30× — 150 mL; 60× — 2 × 150 mL).

6. Incubate at 30°C and 200 rpm until the cell titer is at least 2×10^7/mL. This will take at least 4 hours.

7. Boil the SS carrier DNA (30× — 2.0 mL; 60× — 3.5 mL) for 5 minutes and chill in ice water.

8. Make up appropriate volumes of transformation mixture and keep in ice water.

Ingredients	30× Scale up (mL)	60× Scale up (mL)
PEG 3350 (50% wt/vol)	7.2	14.4
LiAc 1.0 M	1.08	2.16
SS carrier DNA (2 mg/mL)	1.50	3.0
Plasmid DNA + water	1.02	2.04
Total volume	10.8	21.6

9. Harvest the cells by centrifugation, wash them in one half volume of sterile water, and resuspend them in 10 mL of sterile water. Transfer to a 50-mL centrifuge tube. Centrifuge to pellet the cells and discard the supernatant.

10. Pipet the transformation mixture onto the cell pellet and suspend the cells by vortex mixing the tube vigorously.

11. Incubate the tube at 42°C for 45 minutes. Mix the contents by inversion at 5-minute intervals to ensure temperature equilibration.

12. Centrifuge at 3000× g for 5 minutes and remove the transformation mixture.

13. Resuspend the cells in sterile water (30× — 20 mL; 60× — 40 mL) and spread 400-μL samples onto 150-mm plates of SC minus selec-

tion agar (30× — 50 plates; 60× — 100 plates) to select for the presence of both plasmids and reporter gene activation.

14. Incubate at 30°C for 4 to 7 days and score transformants for activity of the reporter gene resulting from interaction between the proteins encoded by *YFG* and the cDNA sequences carried by the secondary plasmid.

Electroporation

The protocol set out below is based on that described by Manivasakam and Schiestl (15). The specific conditions have been optimized for our system: a Gene Pulser (Bio-Rad Laboratories, Hercules, CA, USA) connected to a pulse controller. The PEG and carrier DNA solutions are the same as for the LiAc/SS carrier DNA/PEG protocols.

1. Inoculate the yeast strain into 100 mL of YPAD broth in a 250-mL flask and incubate overnight at 200 rpm and 30°C.

2. Place a tube of carrier DNA (2.0 mg/mL) in a boiling water bath for 5 minutes and then chill in ice water.

3. Harvest the cells by centrifugation in two 50-mL centrifuge tubes, combine the cell pellets, and wash three times in 45 mL sterile distilled/deionized water.

4. Determine the cell titer, centrifuge, and resuspend the cells in water at 1×10^9 cells/mL.

5. Dispense 500-μL samples of the suspension (5×10^8 cells) into 1.5-mL microfuge tubes, pellet the cells, and remove the supernatant.

6. Make up an appropriate volume of the electroporation mixture listed below.

Component	Volume (μL)
PEG 3500 (50% wt/vol)	260
SS carrier DNA (2.0 mg/mL)	37.5
Plasmid DNA (20 ng/μL)	2.5
Water (distilled/deionized)	10.0
Total volume (including cells)	310

7. Pipet 310 μL of electroporation mixture onto the cell pellet and resuspend the cells by vortex mixing vigorously.

8. Incubate the tubes at 42°C for 40 minutes.

9. Vortex mix the tubes and pipet the cell suspension into a chilled 0.2-cm electroporation cuvette.

10. Electroporate at 1.0 kV, 25 μF and 200 Ω.

11. Immediately add 640 μL ice-cold 1.0 M sorbitol to the cuvette.

12. Mix and plate 100-μL samples onto plates of SC minus selection agar.

Spheroplast Transformation (after Spencer et al. [21])

The efficiency of this protocol is very dependent on the activity of the zymolyase used to produce spheroplasts. It is advisable to test a range of zymolyase concentrations with a constant number of cells to determine the optimum enzyme concentration for recovery of transformants (21).

1. Inoculate the yeast strain into 5 mL of YPAD broth in a 15-mL culture tube and incubate overnight at 200 rpm and 30°C.

2. Determine the cell titer. About 4×10^8 cells will be required.

3. Pipet 50 mL of warm YPAD broth into the sterile flask and add 3.75×10^8 cells to give a final titer of 7.5×10^6 cells/mL.

4. Incubate the flask on a rotary shaker at 200 rpm for 4 hours. This will allow the cells to undergo two cell divisions to about 3×10^7 cells/mL and a total of about 1.5×10^9 cells.

5. Harvest the cells by centrifugation and wash them once in 20 mL sterile water and once in 20 mL 1 M sorbitol. Centrifuge again and discard the supernatant.

6. Resuspend the cells in 20 mL of SPEM (1 M sorbitol, 10 mM sodium phosphate, pH 7.5, 10 mM EDTA plus 40 μL of β-mercaptoethanol added immediately before use).

7. Add 45 μL of zymolyase 20T (10 μg/mL) and incubate at 30°C for 20 to 30 minutes with gentle shaking.

8. Assay for spheroplast formation by adding 10 μL of the reaction to a drop of 5% SDS on a microscope slide and examine with phase contrast microscopy. Continue if 80% to 90% of the cells appear as ghosts. Titrate the activity of the zymolyase if spheroplasting takes longer than 40 minutes.

9. Centrifuge the spheroplast preparation for 4 minutes at 250× g; remove the supernatant carefully and rinse the pellet once in 20 mL of STC (1 M sorbitol, 10 mM Tris-HCl, pH 7.5, 10 mM $CaCl_2$) and resuspend in 2 mL of STC.

10. Pipet 150-μL samples of the spheroplast preparation into 15-mL tubes and add up to 5 μg of plasmid DNA in a total volume of 10 μL. Mix gently and incubate at room temperature for 10 minutes.

11. Add 1 mL of PEG 8000 solution (PEG MW 8000 20% wt/vol, 10

mM Tris-HCl, pH 7.5, 10 mM $CaCl_2$) and mix gently by inversion. Incubate at room temperature for 10 minutes.

12. Centrifuge at 250× *g* for 5 minutes and remove the supernatant. Resuspend the spheroplasts in 150 μL of SOS (see section entitled Media for components of medium) and incubate at 30°C for 30 minutes.

13. Add 6 mL of regeneration agar equilibrated to 46°C, mix by inversion, and pour as an overlay onto plates of SC (Sorb) minus selection agar.

14. Incubate the plates at 30°C for 3 to 4 days.

MEDIA

YPAD and 2× YPAD Media

Many commonly used yeast strains are mutant for the *ade2* gene and accumulate a red pigment when grown on YPD (yeast extract peptone dextrose) medium. These strains grow better and do not accumulate the red pigment on YPAD medium (YPD supplemented with adenine at 100 mg/L). We have found that cultures grow faster and give higher transformation rates if they are regrown for two divisions in 2× YPAD broth (double strength YPAD) prior to transformation by the high-efficiency protocol. We mix all media in volumes of 600 or 800 mL. These volumes are easier to handle than 1000 mL and can be made up and autoclaved in 1.0-L Pyrex® brand medium bottles. The ingredients for 800-mL samples of YPAD agar and 2× YPAD broth are as follows:

Ingredients	YPAD Agar	2× YPAD Broth
Bacto Yeast Extract	8 g	16 g
Bacto Peptone	16 g	32 g
Glucose	16 g	32 g
Adenine hemisulphate	80 mg	80 mg
Bacto Agar (omit the agar to make YPAD broth)	12 g	—
Distilled/deionized water	800 mL	800 mL

Add the powdered ingredients to the water on a stirring hot plate with the temperature set to high. Continue stirring and allow the YPAD agar to boil for 1 minute to ensure that the agar is dissolved. Sterilize by autoclaving for 15 minutes, and allow the medium to equilibrate to 56°C in a water bath before pouring into petri dishes. This volume of medium is sufficient for about 30 plates. YPAD and 2× YPAD broth are dissolved on a stirring hot plate, dispensed in 100- or 200-mL aliquots and autoclaved. Commercial formulations of YPD agar and YPD broth are available and may be more convenient than making media from scratch.

SC Minus and SC (Sorb) Minus Media

Transformants are selected on plates of SC minus agar (synthetic complete medium) (17) appropriate to the selectable yeast gene carried by the plasmid used for transformation. The most commonly used selectable markers are *URA3* (uracil requirement), *TRP1* (tryptophan requirement), *HIS3* (histidine requirement), *LEU2* (leucine requirement), and *ADE2* (adenine requirement). These compounds are omitted singly or in combination from the amino acid mixture listed to select for specific plasmids.

SC minus agar and SC (Sorb) minus agar for spheroplast transformation contain:

Ingredient	SC minus agar	SC(Sorb) minus agar
Yeast Nitrogen Base without amino acids	5.4 g	5.4 g
Amino acid mixture [the mixture should lack the nutrient(s) corresponding to the selected marker(s)]	1.6 g	1.6 g
Sorbitol	—	145.6 g
Glucose	16.0 g	16.0 g
Bacto agar (omit the agar for SC minus liquid medium)	12.0 g	12.0 g
Distilled/deionized Water	800.0 mL	800.0 mL

The ingredients should be added to the water on a cold stirring hot plate, and the pH adjusted to 5.6 with 1.0 N NaOH. Turn the temperature to high, continue stirring, and boil the medium for 1 minute to ensure that the agar is dissolved. The medium is then autoclaved for 15 minutes and allowed to equilibrate to 56°C in a water bath before it is poured into petri dishes. The plates are allowed to dry in the dark at room temperature for 1 or 2 days and then stored in sealed bags in the dark at 4°C.

The amino acid mixture contains:

Adenine SO$_4$	0.5 g	Methionine	2.0 g
Arginine	2.0 g	Phenylalanine	2.0 g
Aspartic Acid	2.0 g	Serine	2.0 g

Glutamic Acid	2.0 g	Threonine	2.0 g
Histidine HCl	2.0 g	Tryptophan	2.0 g
Inositol	2.0 g	Tyrosine	2.0 g
Isoleucine	2.0 g	Uracil	2.0 g
Leucine	4.0 g	Valine	2.0 g
Lysine HCl	2.0 g	p-aminobenzoic acid	0.2 g

The ingredients are placed into a plastic container and thoroughly mixed by adding several glass marbles and shaking the container for 5 minutes.

SOS Medium and Regeneration Agar for Spheroplast Transformation (21)

SOS medium contains:

Sorbitol	91.0 g
$CaCl_2 \cdot 2H_2O$	0.5 g
Yeast extract	1.25 g
Peptone	2.5 g
Amino acid mixture [the mixture should lack the nutrient(s) corresponding to the selected marker(s)]	1.0 g
Distilled/deionized water to	500 mL

Regeneration agar is made to the same formula as SC (Sorb) minus medium but with 20 g Bacto Agar (2.5% wt/vol).

Since regeneration agar and SOS are used in small quantities, they should be dispensed into smaller containers before or after autoclaving. Regeneration agar must be melted and equilibrated to 46°C before it is used.

SOLUTIONS

Single-Stranded Carrier DNA (2.0 mg/mL)

Dissolve 200 mg of salmon sperm DNA (Cat. No. D-1626; Sigma) in 100 mL of TE buffer (10 mM Tris-HCl, 1 mM Na_2 EDTA, pH 8.0) on a stir plate overnight at 4°C. Dispense 1.0-mL samples into 1.5-mL microcentrifuge tubes and store at -20°C. The carrier DNA is denatured by boiling for 5 minutes before use. A sample of carrier DNA can be boiled 3 or 4 times without loss of activity.

PEG MW 3350 (50% wt/vol)

Dissolve 50 g of PEG 3350 (Cat. No. P-3640; Sigma) in 30 mL of distilled/deionized water in a 150-mL beaker in a water bath at 56°C. Pour into a 100 mL measuring cylinder, rinse out the beaker with water, and add to the solution in the cylinder. Cool to room temperature and add water to make the volume up to 100 mL. Seal the cylinder with parafilm and mix by inversion. Pour the PEG solution into a glass bottle and autoclave for 15 minutes. Store securely capped at room temperature. This solution is for the LiAc/SS carrier DNA/PEG and electroporation protocols.

PEG MW 8000 (20% wt/vol)

Dissolve 20 g of PEG 8000 (Cat. No. P-4468; Sigma) in 30 mL of distilled/deionized water and proceed as for PEG 3350. This solution is for the spheroplast transformation protocol.

Lithium acetate (1.0 M)

Dissolve 5.1 g of lithium acetate dihydrate (Cat. No. L-6883; Sigma) in 50 mL of water. Sterilize by autoclaving for 15 minutes and store at room temperature.

TROUBLESHOOTING

LiAc/SS Carrier DNA/PEG Transformation

The most likely reason for poor transformation is a refractory yeast strain. For example, the strain CG378 transforms 200 times more efficiently than the strain GGY1::171 (19), and the strain Y190 (6) transforms well by the high-efficiency protocol, but very poorly by the rapid transformation protocol (R.A. Woods, unpublished data). It is best to use strains that are reported as good transformers in the literature. Temperature-sensitive mutant strains may not survive incubation at 42°C. They can be incubated overnight at 30°C in transformation mixture and then plated onto SC minus selection agar.

It is most important to store the PEG in a bottle with a tight cap. The optimum concentration of PEG in the transformation reaction is 33%, and evaporation of the PEG solution during storage will result in poor transformation. Some protocols suggest that cells should be suspended in TE after transformation. We have found that this inhibits growth to some degree. It is also important to adjust the pH of the SC minus medium to 5.6 before it is autoclaved; higher and lower values reduce the number of transformants.

Electroporation

Yeast strains also vary in their susceptibility to electroporation. If a strain responds poorly, then investigate the effects of voltage, resistance, concentration of plasmid, electroporation buffer, and postelectroporation treatment. Alternatively, use the LiAc/SS carrier DNA/PEG protocols.

Spheroplast Transformation

The most critical phase of this protocol is the conversion of the cells to spheroplasts. If 80% to 90% of the cells are not spheroplasted in 20 to 30 minutes, then the zymolyase should be titrated and a suitable concentration determined. The spheroplasts are very fragile; they should be centrifuged at no more than $300 \times g$ and pipetted and resuspended very gently. For the same reason, the regeneration agar should be equilibrated to 46°C before it is added to the transformed spheroplasts.

SUPPLIERS

Becton Dickinson Microbiology Systems, Cockeysville, MD, USA
   BBL YEPD Agar and YEPD Broth Media.

Bio-Rad Laboratories, Hercules, CA, USA
   Gene Pulser and Pulse Controller.

Corning, Science Products Division, Corning, NY, USA
   Pyrex brand media bottles. A wide range of sizes with pouring lips
   and melamine caps.

Difco Laboratories, Detroit, MI, USA
   Bacto Yeast Extract, Bacto Peptone, Bacto Agar, and Yeast Nitrogen
   Base.

International Equipment Company (IEC), Needham Heights, MA, USA,
   IEC centrifuges.

ICN Biomedical Research Products, Costa Mesa, CA, USA
   Zymolyase 20T.

Sigma, St. Louis, MO, USA
   Lithium acetate, PEG 3350, PEG 8000, and salmon sperm DNA.

Bio/Can Scientific, Mississauga, Ontario, Canada or Tetra Link International, Amherst, NY, USA.
   Molecular Research Reagents Inc. (MRRI) Gietz Lab Yeast Transformation Kit containing PEG, LiAc, Carrier DNA, a control plasmid, and complete instructions. Cat. No. MRR-GYT-1211.

ACKNOWLEDGMENTS

This work was supported in part by grants from the Medical Research Council of Canada to R.D.G. and from the University of Winnipeg to R.A.W. We should also like to acknowledge Dr. Angela Dyer, then a University of Winnipeg student, who optimized the electroporation protocol for our laboratory.

REFERENCES

1. Becker, D.M. and L. Guarente. 1991. High-efficiency transformation of yeast by electroporation. Methods Enzymol. *194*:182-187.
2. Beggs, J.D. 1978. Transformation of yeast by a replicating hybrid plasmid. Nature *275*:104-109.
3. Brzobohaty, B. and L. Kovac. 1986. Factors enhancing genetic transformation of intact yeast cells modify cell wall porosity. J. Gen. Microbiol. *132*:3089-3093.
4. Burgers, P.M.J. and K.J. Percival. 1987. Transformation of yeast spheroplasts without cell fusion. Anal. Biochem. *163*:391-397.
5. Chien, C., P.L. Bartel, and S. Fields. 1991. The two-hybrid system: a method to identify and clone genes for proteins that interact with a protein of interest. Proc. Natl. Acad. Sci. USA *88*:9578-9582.
6. Durfee, T., K. Becherer, R.-L. Chen, S.H. Yeh, Y. Yang, A.K. Kilburn, W.H. Lee, and S.J. Elledge. 1993. The retinoblastoma protein associates with the protein phosphatase type 1 catalytic subunit. Genes Dev. *7*:205-214.
7. Gietz, R.D. and A. Sugino. 1988. New yeast-*Escherichia coli* shuttle vectors constructed with *in vitro* mutagenised yeast genes lacking six-base pair restriction sites. Gene *74*:527-534.
8. Gietz, R.D., A. St. Jean, R.H. Schiestl, and R.A. Woods. 1992. Improved method for high efficiency transformation of intact yeast cells. Nucleic Acids Res. *20*:1425.
9. Gietz, R.D. and R.A. Woods. 1994. High efficiency transformation with lithium acetate, p. 121-134. *In* J.R. Johnston (Ed.), Molecular Genetics of Yeast: A Practical Approach. Oxford University Press, Oxford.
10. Gietz, R.D., R.H. Schiestl, A.R. Willems, and R.A. Woods. 1995. Studies on the transformation of intact yeast cells by the LiAc/SS-DNA/PEG procedure. Yeast *11*:355-360.
11. Gietz, R.D., B. Triggs-Raine, A. Robbins, K.C. Graham, and R.A. Woods. 1997. Identification of proteins that interact with a protein of interest: applications of the yeast two-hybrid system. Mol. Cell. Biochem. *172*:67-79.
12. Grey, M. and M. Brendel. 1992. A ten-minute protocol for transforming *Saccharomyces cerevisiae* by electroporation. Curr. Genet. *22*:335-336.
13. Hinnen, A., J.B. Hicks, and G.R. Fink. 1978. Transformation of yeast. Proc. Natl. Acad. Sci. USA *75*:1929-1933.
14. Ito, H., Y. Fukuda, K. Murata, and A. Kimura. 1983. Transformation of intact yeast cells treated with alkali cations. J. Bacteriol. *153*:163-168.
15. Manivasakam, P. and R.H. Schiestl. 1993. High efficiency transformation by electroporation. Nucleic Acids Res. *21*:4414-4415.
16. Potter, H. 1993. Application of electroporation in recombinant DNA technology. Methods Enzymol. *217*:461-478.
17. Rose, M.D. 1987. Isolation of genes by complementation in yeast. Methods Enzymol. *152*:481-504.
18. Schiestl, R.H. and R.D. Gietz. 1989. High efficiency transformation of intact yeast cells using single stranded nucleic acids as carrier. Curr. Genet. *16*:339-346.
19. Schiestl, R.H., P. Manivasakam, R.A. Woods, and R.D. Gietz. 1993. Introducing DNA into yeast. Methods: A Companion to Methods in Enzymology *5*:79-85.
20. Simon, J.R. 1993. Transformation of intact yeast cells by electroporation. Methods Enzymol. *217*:478-483.

21. Spencer, F., G. Ketner, C. Connelly, and P. Hieter. 1993. Targeted recombination-based cloning and manipulation of large DNA segments in yeast. Methods: A Companion to Methods in Enzymology 5:161-175.

22. Struhl, K., D.T. Stinchcomb, S. Scherer, and R.W. Davis. 1979. High-frequency transformation of yeast: autonomous replication of hybrid DNA molecules. Proc. Natl. Acad. Sci. USA 76:1035-1039.

23. Traver, C.N., S. Klapholz, R.W. Hyman, and R.W. Davis. 1989. Rapid screening of a human genomic library in yeast artificial chromosomes for single-copy sequences. Proc. Natl. Acad. Sci. USA 86:5898-5902.

4 | Transformation of *Dictyostelium*

David A. Knecht[1] and Eunkyung Lee[2]
[1]University of Connecticut, Department of Molecular and Cell Biology, Storrs, CT; [2]Department of Cell Biology, Yale University School of Medicine, Buyer Center for Molecular Medicine, New Haven, CT, USA

INTRODUCTION

Dictyostelium discoideum is a eukaryotic soil amoeba that has been a popular model system for the study of cell physiology and developmental biology since its characterization in the 1940s (33). Amoebas normally use bacteria as a food source, but axenic strains have been developed that can also be grown in a simple broth medium. Much of the initial interest in this system stemmed from its unusual and simple developmental program in which a population of identical, free living amoeba gather together into groups of 100 to 10^6 cells to form a multicellular organism. A great deal is now understood about the mechanisms of cell type differentiation and morphogenesis of this organism (22). More recently, the similarity of cell structure and behavior to mammalian cells has made the genetic approaches available appealing for studying basic cell biological problems like motility, phagocytosis, and signal transduction.

Cells can be maintained in the laboratory as either stable haploids or diploids. Most investigation has focused on the haploid state, since mutational alterations are simpler to interpret. The lack of classical meiotic genetics has made the investigation of mutants more difficult; however, the development of a transformation system for *Dictyostelium* has allowed the development of a variety of molecular genetic alternatives. It is now relatively simple to introduce genes into amoeba, as well as to create mutations that alter endogenous genes through homologous gene targeting. The ongoing *Dictyostelium* genome project should provide a wealth of new tools for investigating and manipulating the *Dictyostelium* genes (23).

Gene Transfer Methods
Edited by Pamela A. Norton and Laura F. Steel
©2000 Eaton Publishing, Natick, MA

Transformation of *Dictyostelium* became possible with the development of a vector that could be reliably used to select for drug-resistant amoeba (28). The system currently in use is quite similar to that developed for mammalian cells, which is an indication of the similarity of the cell architecture and behavior (6,31). The original *Dictyostelium* transformation vector contained the bacterial *TN5* gene fused to a *Dictyostelium* actin 6 promoter. This *TN5* gene product detoxifies both kanamycin (which kills bacteria, but not eukaryotic cells) and G418 (which efficiently kills eukaryotic cells). G418 was found to reliably kill untransformed amoeba at concentrations 100 times lower than what is required to kill mammalian cells. Subsequent vectors added a *Dictyostelium* termination sequence after the *TN5* gene, and then variants were constructed with the actin 15 promoter and the TN903 resistance gene (18). The introduction of these plasmids into *Dictyostelium* amoeba also closely followed the methods used for transfecting mammalian cells. DNA was calcium phosphate precipitated, and then allowed to settle onto a monolayer of amoeba. The only modification in adapting the mammalian protocol for *Dictyostelium* was to use HL5 growth medium instead of tissue culture medium. Cells were then glycerol-shocked and allowed to recover in growth medium. Early protocols for selecting resistant cells were complex, but eventually it was realized that transformed cells would grow in broth media on a plastic petri dish surface just like mammalian cells.

TRANSFORMATION TECHNIQUES

Dictyostelium cells can be transformed using several different techniques. The original method is treating cells with $CaPO_4$-precipitated DNA. Cells become drug resistant with a frequency of about 1 in 10^6 using an integrating G418 resistance vector. Use of an extrachromosomal vector (see below) increases this efficiency by 10- to 100-fold. An alternative to $CaPO_4$ coprecipitation is electroporation. This technique works very efficiently and reproducibly for transforming amoeba and is simpler and faster to carry out. Lipid reagents also work for transforming cells (W. Nellen, personal communication), but have not been used extensively.

COPY NUMBER AND VECTOR INTEGRATION SITE

Transformation of *Dictyostelium* has become routine, and a variety of vectors have been constructed that provide a wide range of options for introducing DNA into cells. Most of these vectors are *Escherichia coli*-based plasmids with a *Dictyostelium*-selectable marker added. These vectors seem to insert randomly into the genome, in that they can be found in a variety of chromosomal locations (7). Each clonal transformant appears to have a single integration site. With G418-based selections, there is usually a tandem array of several hundred copies of the vector at a single integration site. This observation led to the belief that cells need a large number of copies of the resistance gene in order to produce sufficient neomycin phosphotransferase to inactivate the drug. However, in subsequent homologous targeting experiments (discussed below), cells containing only a single copy of the vector were isolated. It is still unclear what determines the copy number of the vector. Recent results from our laboratory have shown that there is a consistent trend in copy number and expression level depending upon the drug selection used. Cells selected for G418 resistance consistently have a higher vector copy number and a higher level of expression of a marker gene (green fluorescent protein; GFP) present in the plasmid. Blastocidin-resistant cells have a very low vector copy number and expression level, but the efficiency of transformation with these vectors is higher than with G418. Hygromycin-resistant cells have a wide range of copy number and expression level (30), but for reasons that are not understood, the efficiency of transformation with these vectors is extremely low.

Cotransformation has been shown to occur at high efficiency in *Dictyostelium* (27). If two vectors are mixed and used to transform cells, about 50% of the cells selected for the presence of one of the vectors will also contain the other vector. The pattern of the integration indicates that vectors undergo homologous recombination with each other in virtually all cases. In addition, linearizing the plasmid had no effect on the recombination site and was in all cases repaired (17). This is consistent with results indicating that linearizing vector DNA has no effect on transformation efficiency (D.A. Knecht, unpublished results).

The isolation of several endogenous plasmids from wild strains of *Dictyostelium* allowed the construction of stable extrachromosomally replicating vectors by adding a drug resistance marker to the plasmid (25,29). These vectors are maintained extrachromosomally at several hundred copies per cell and are quite stable in the absence of selection. The plasmids can be quite large, and so constructs based upon them can be difficult to make. To circumvent this problem, Manstein and coworkers (24) recently developed a dual plasmid transfection system. One vector contains a *Dictyostelium* origin of replication from the extrachromosomal plasmid (Ddp2 ori), a selectable marker, and an expression cassette where the gene of interest can be inserted for expression under control of the constitutive actin (pDXA) or inducible discoidin (pDXD) promoter. The second plasmid (pREP) contains the open reading frame (ORF) gene from the endogenous plasmid Ddp2. When cells are cotransformed with the two plasmids, the pREP plasmid provides the ORF gene function *in trans* to the plasmid containing the Ddp2 ori, thereby maintaining it as an episome. If the pDXA vectors are used without the complementing plasmid, they integrate into the genome.

DRUG AND NUTRITIONAL SELECTIONS

A number of new selection systems have also been developed, allowing multiple vectors to be sequentially introduced into a single cell. In addition to G418 resistance (neo), vectors that encode resistance to hygromycin (14), blasticidin (Bsr) (40), and bleomycin (4) are available. Several nutritional selections have also been developed. Mutants lacking the thymidylate synthase gene require thymidine supplementation of normal growth medium (HL5) for growth. Vectors containing either the *Dictyostelium* or mouse versions of the thymidylate synthase gene (*thy*) have been constructed and can be used to select for transformants that grow in normal unsupplemented medium (5,13). The only disadvantage to this system is that one must start with the available thy⁻ cell lines. There is no selection for thy⁻ cells, so isolating new ones requires a time-consuming mutant screen.

A more robust auxotrophic selection system based upon the *ura* gene encoding the enzyme UMP-synthase has been developed by Dingermann and coworkers (16). Normal cells containing the *pyr56* gene are killed when grown in medium containing 5-fluoro-orotic acid (5-FOA). This compound becomes converted to 5-fluoro-UMP by UMP synthase, and then binds irreversibly to thymidylate synthase, leading to cell death. These workers used a vector containing a portion of the *pyr56* gene to transform wild-type cells and selected for 5-FOA resistance. Since the fragment of the gene was not mutated and not under direct selection, the presumption is that it would cause a targeted gene conversion or defective integration event which would inactivate the *pyr56* gene. In fact, this approach worked, and ura⁻ cells can be isolated in this way. This selection is not very efficient and made difficult by the fact that cells treated with 5-FOA become blocked in the cell cycle and as a result begin to aggregate. Therefore, the selection is best done at a very low cell density to minimize the cells' ability to signal each other to begin aggregation. These ura⁻ cells require uracil supplementation of HL5 medium for growth. In the second step, a vector containing the *pyr56* gene and a gene of interest can be introduced into the ura⁻ cells by selecting for cells that can grow in unsupplemented medium. In theory, these cells can again be rendered ura⁻ by 5-FOA selection and the process repeated to allow for multiple rounds of introduction of plasmids into the same cell line.

APPLICATIONS

Expression

Transformation of *Dictyostelium* cells has been used to accomplish a variety of aims. Expression of proteins in transformed cells has been used for functional analysis of genes as well as for purification of correctly folded and modified proteins. Several vectors have been constructed for this purpose by adding a second expression cassette to the basic G418 resistance vectors. This has been done using actin 6, actin 15, or discoidin promoters. All three promoters are expressed during growth in axenic medium and in early development. The actin 15 promoter is turned off during vegetative growth on bacteria (H. MacWilliams, personal communication). Thus, cells can be induced by switching from growth on bacteria to growth in HL5 or by plating for development. The discoidin promoter can be repressed during vegetative growth at low density in HL5 by adding 1 mM folic acid to the medium. Induction is accomplished by changing to medium with no folic acid, and allowing cells to grow to high density, or by replacing medium with conditioned medium from high titer cells (2).

A particularly useful series of expression vectors has been constructed by Manstein et al. (24). These vectors include either actin or discoidin promoters and either His and myc affinity tags adjacent to the polylinker site. These vectors (and others) have been used to express both endogenous and exogenous proteins in *Dictyostelium* and to subsequently purify the protein from the cells.

There was some initial concern that, with the unusually high AT bias of the *Dictyostelium* genome and the preference of AT at third codon positions used by endogenous genes, there might be a problem with expressing foreign genes in *Dictyostelium*. However, genes from bacteria, yeast, mammals, plants, and fungi have all been successfully expressed in *Dictyostelium* without any modification of the codons indicating that this is not a general problem. There is some evidence that, for foreign secreted proteins, changing the first few amino acids to *Dictyostelium* codons enhances expression (A. De Lozanne, personal communication).

Homologous Gene Targeting

Transformation vectors can be used to target vectors to specific chromosomal sites by homologous recombination. In this way, gene disruptions, gene deletions, and gene modifications have been generated. The events that occur are to some extent dependent upon the vector used. Single cross-over gene disruptions are typically created by taking a fragment of the gene of interest and inserting it into a plasmid containing a selectable marker. When transformed into cells as a supercoiled molecule, this vector will integrate at the homologous site by a single cross-over recombination event at a frequency that is variable, but approaches 10% to 30% (Figure 1). If the gene fragment does not contain the 3′ end, this single cross-over event will lead to a disrup-

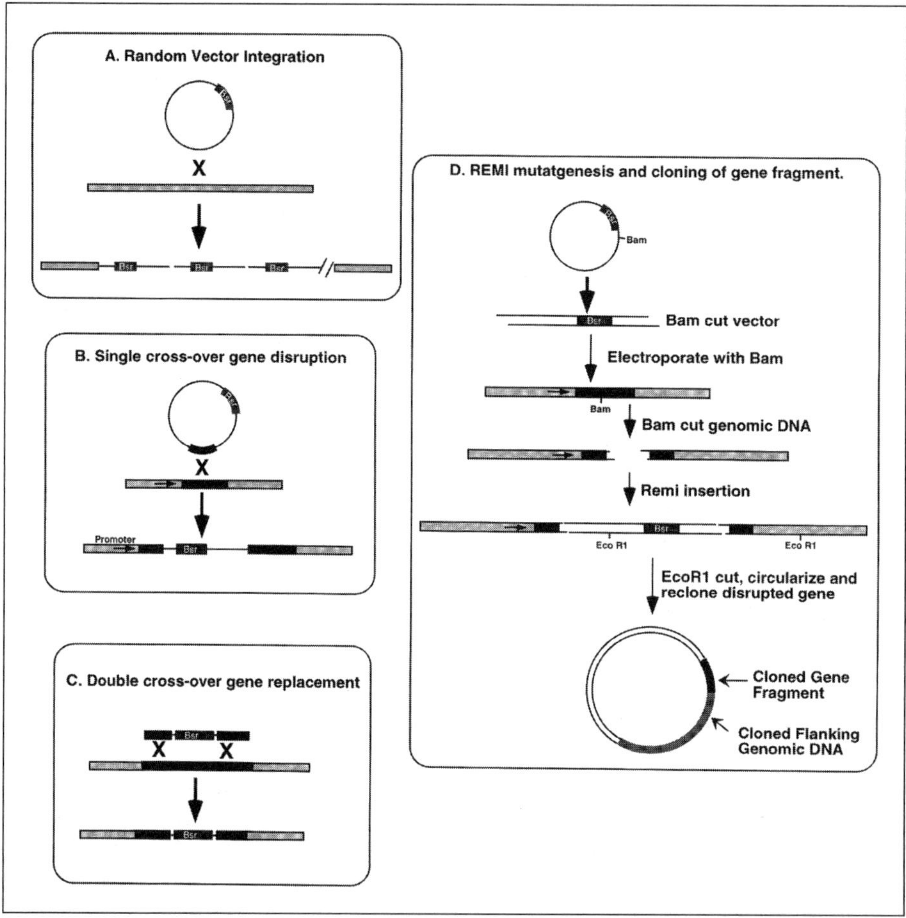

Figure 1. Alternative outcomes of vector insertion into the genome. (A) If a circular plasmid is introduced into cells, it will integrate randomly at a single site as a tandem array. The number of copies depends on the vector and drug selection used. (B) If the vector contains sequences homologous to chromosomal sequences, integration will occur at the targeted locus as a tandem array in a small proportion of the transformed cells (the rest will have random integration). (C) If a linear vector with flanking homologous sequences is used, the vector will replace the chromosomal copy in a small proportion of the transformed cells. (D) REMI mutagenesis and mutant gene isolation: a REMI vector is linearized with a single cut restriction enzyme (*Bam*H1 is shown). This linear DNA is electroporated into cells along with *Bam*H1 (or *Dpn*II). Transformants are screened for mutations caused by integration of the vector at chromosomal *Bam*H1 sites. The disrupted gene is isolated by cutting the genomic DNA from the mutant with a restriction enzyme that releases the vector along with part of the disrupted gene. This DNA is circularized and then used to transform *E. coli* to ampicillin resistance.

tion of the endogenous gene at the distal end of the targeting sequence. If the portion of the gene that is still made is nonfunctional, then a null mutant will result. Since extra copies of the vector have no effect on the disruption event, vectors containing a G418 selection that integrate with several hundred copies at the insertion site have been used for these experiments (e.g., see References 8 and 10). Gene replacement or gene deletions, on the other hand, require that a single copy of the vector replace the endogenous gene through a double cross-over recombination event (Figure 1). In this case, blasticidin vectors, which typically integrate with 1 to 5 copies per cell, have proved to be more efficient than G418 vectors. Hygromycin vectors have recently been shown to give a range of copy number and expression levels, and so are theoretically suitable for either type of experiment (30). The nutritional selections for *thy* and *ura* genes require only a single copy for selection, and so they have also been used extensively for gene knock-out experiments.

Several approaches have been used to increase the frequency of homologous gene targeting events. For double cross-over events to occur, the DNA is introduced into the cell as a linear fragment. *Dictyostelium* can efficiently repair those ends and religate them. In order to prevent this, Ratner and coworkers (38) treated the ends with terminal transferase in the presence of 2'3'dideoxynucleotides. This created an end that could not be ligated and presumably could not be efficiently repaired in the cell, and as a result, increased the targeting efficiency. Morrison et al. (26) made use of a positive-negative selection to try to increase the targeting efficiency. In mammalian cells, this is usually accomplished by placing the *neo* gene between the homologous targeting sequences and the thymidine kinase gene (*tk*) in the flanking plasmid sequence. Selection for *neo* and against *tk* selects for double cross-over recombination events in which the flanking vector DNA is lost. In *Dictyostelium*, the *tk* selection does not work; however, a lethal gene was created by mutating a tRNA gene to suppress the stop codon UAA (12). Since nearly every *Dictyostelium* gene uses this stop codon, its presence is lethal. By introducing a UAA suppressor tRNA into the flanking vector, the targeting efficiency was increased to over 90%. However, many of these events were not simple double cross-overs. Ratner found that to increase the frequency of simple targeting events, both the lethal tRNA gene and the dideoxynucleotide treatment was necessary (38).

Gene Identification and Isolation

An important application of transformation technology is gene complementation. A mutation is isolated by chemical or UV mutagenesis and screening or selecting for a phenotype. The mutant gene is then isolated by introducing a library of wild-type genes into the mutant cell, and screening for reversion of the mutation. This methodology is particularly well established in yeast and bacteria. Although it has been accomplished in *Dictyostelium* (13), gene complementation has not been routinely repeatable for unknown reasons. A recent report of success with complementation may lead to renewed interest in this approach (D. Robinson, personal communication).

An alternative approach to mutant isolation, which has been very successful, is the use of restriction enzyme-mediated integration (REMI) (Reference 20 and Figure 1). In this technique, a linearized transformation vector (usually a blasticidin-based plasmid) is electroporated into cells along with a restriction enzyme that cuts at a ligation compatible site (e.g., plasmid cut with *Bam*H1 and electroporation with *Sau*3A1 or *Dpn*II). Apparently, the enzyme cuts the genomic DNA at specific sites, and the vector DNA, with compatible ends, becomes ligated into the cut site. Thus, it is possible to generate a library of mutants created by integration of the vector into each restriction enzyme site in the genome. In some cases, cells are transformed with the REMI vector to generate a population of transformants, and then these cells are put through a selection to identify a particular trait. In other cases, the cells are cloned after transformation, and cell lines are screened individually for defects. Many mutants have been identified by screening for REMI transformants that do not develop normally. Once a mutant is identified, the gene that was disrupted by the vector is recloned from the mutant cell. In order to do this, genomic DNA from the mutant strain is digested with a restriction enzyme that cuts once in the vector sequence. This will cleave the vector into two pieces, but one piece will contain the bacterial origin and *amp* gene, as well as a portion of the disrupted gene up to the next restriction site. When the DNA is religated into circles and then used to transform *E. coli*, any cells able to grow contain a piece of the gene of interest. Many mutants have already been characterized by this approach and now the technology is being extended to isolate second site suppressors in mutant strains.

PROTOCOLS

Components and concentrations of all solutions, buffers, and media are detailed in the section entitled Stock Solutions.

Transformation

Electroporation

Developed by Eunkyung Lee as a modification of Schlatterer et al. (36), this is the recommended protocol because it gives the greatest efficiency and is relatively simple.

1. Pellet approximately 5×10^6 cells by centrifugation at 1500 rpm (about $500\times g$) for 5 minutes at 4°C.

2. Wash the cells 1 to 2× with ice-cold H-50 buffer.

3. Resuspend the pellet with 100 μL H-50 buffer.

4. Add DNA in Tris-EDTA (TE) or water to the cell suspension. We usually use 3 to 4 μg, but more DNA will lead to more transformants. Try to use as small a volume of DNA as possible to not alter the composition of the electroporation buffer.

5. Transfer the cells to a cold 0.1-cm electroporation cuvette.

6. Electroporate the cells at 0.85 kV/25 μF (time constant is approximately 0.6 msec) twice with 5 seconds between pulses.

7. Leave the cells in the cuvette and incubate on ice for 5 minutes.

8. Remove as much volume as possible from the cuvette with a pipettor or pasteur pipet. Add HL5 medium to the cuvette to recover the remaining cells and plate in a 100-mm petri dish with a total of 10 mL of HL5 medium. Alternatively, cells can be aliquoted into wells of a 96-well microplate at this time (about 75 μL/well).

9. Add the drug for selection the next day. (G418, 10 μg/mL; hygromycin, 25 μg/mL; Blasticidin, 4 to 10 μg/mL final concentration).

In 2 to 3 days, the dead untransformed cells in the petri dishes round up and detach from the plate. At this point, gently swirl the plate to resuspend the dead cells and aspirate the medium. The medium is replaced with fresh HL5 containing selective drug. The plates are left until the transformants have grown significantly (up to a week), and then the medium is changed again to remove the rest of the dead cells. Often, some of the initial colonies that appear will die in that first week, but some cells do survive, so be patient. Some strains, like NC4A2 and AX3, are very motile during growth, and so colonies, if they form at all, will contain very dispersed cells. Other strains, like AX4, form very tight colonies that are easily seen with the naked eye.

Transformation by CaPO$_4$ Coprecipitation

1. The day before transforming (or the morning of) 5×10^6 amoeba growing in HL5 medium at 1×10^6 to 10^7 cells/mL are inoculated into 100-mm Petri dishes containing 10 mL Bis-Tris buffered HL5 (pH 7.1). Bis-Tris is used because it is inexpensive and has an appropriate pK.

2. The next morning (or later that day), change the medium by aspirating it off and replacing it with 10 mL fresh Bis-Tris HL5, pH 7.1, to further reduce the phosphate present in HL5 growth medium. The cells stick

well to the plastic surface and are not dislodged when the medium is gently removed by aspiration.

3. Prepare the DNA: A calcium phosphate precipitate of the DNA to be transformed into the cells is made in a 12- × 75-mm polystyrene test-tube by adding (in order) to a final volume of 1.0 mL: *(i)* 0.4 mL sterile water, *(ii)* 10 µg DNA (about 10 µL of a 1 mg/mL solution), *(iii)* 100 µL 1.25 M $CaCl_2$, and *(iv)* 0.5 mL 2× HBS. No mixing is necessary. Let stand at room temperature for 30 minutes to allow precipitate to form. What you will see is a very slight cloudiness to the solution. If allowed to precipitate too long, the precipitate will settle out.

4. Add the DNA-$CaPO_4$ dropwise directly to the medium, dispersing it around the plate. Let stand several (4–12) hours.

5. Glycerol shock: Carefully aspirate off the medium and gently add 4 mL of 15% glycerol in 1× HBS (just before use, mix equal volume of 2× HBS with 30% glycerol in water solution) by letting it run down the side of the dish. Let stand precisely 5 minutes. Gently aspirate off the glycerol solution and replace with 10 mL regular HL5. Incubate overnight for phenotypic expression.

6. Selection of transformants: Aspirate off the medium and replace with 10 mL fresh HL5 containing appropriate medium and selective agent.

Drug Selections

G418 Selection

This works reliably for electroporation and $CaPO_4$ coprecipitation. For most strains, a concentration of 10 to 20 µg/mL G418 is adequate to kill untransformed cells. The stock solution is made up according to mass rather than activity. In our hands, cells treated with 10 µg/mL G418 look normal after one day, begin to round on day 2, and become detached by day 3. When a new batch of drug is received, it should be tested to see if treated cells follow this pattern. If not, the drug concentration can be adjusted accordingly. AX3 seem to be more resistant to G418 than most strains, and we periodically have background colonies on the control plates. If this happens, the concentration of drug is increased in 2-fold increments until this problem disappears. Cells can be selected at higher concentrations of G418. In general, as the drug concentration is increased in 5-fold increments, the number of transformants decreases, but the expression from the plasmid is increased. Cells have been selected to grow in up to 500 µg/mL G418.

Hygromycin Selection

As above, except use about 25 to 50 µg/mL hygromycin. The drug concentration should be titrated as above for detachment of dead cells at day 3. This selection is very inefficient for $CaPO_4$ coprecipitation, but works reliably for electroporated cells.

Blasticidin Selection

As above, except use 4 to 10 µg/mL blasticidin. Blasticidin does not kill cells very efficiently at high cell density, so cells should be diluted to about 1×10^5 cells/mL or less for selection.

Thy Selection

Start with the thy⁻ parental strain JH10 (13). These cells must be grown in medium supplemented with 40 µg/mL thymidine. The day after transformation, the medium is aspirated, and normal HL5 is added. This selection takes longer than the drug selections above. It is a good idea to do a parallel control with untransformed cells to be sure when the thy⁻ cells are dead. The thy⁺ cells tend to slowly replace the thy⁻ over time, so it is not obvious when the transformants have taken over the population.

FOA/Ura Selection

Several ura⁻ strains are available including DelF-16-11 (16) generated in an AX2 background and DH1 developed in an AX3 background. Generating ura⁻ cells from any parental strain is possible, although not trivial. Cells are transformed with the vector pDU3B1 (16) by electroporation or $CaPO_4$ as indicated above. This vector contains a wild-type copy of the *pyr56* gene, and no one has constructed a true gene deletion construct. Presumably, this vector works by causing a defective recombination or gene conversion that destroys gene function. Since the residual Pyr56 activity in the cells should kill even the cells in which the *pyr56* gene has been inactivated, it is important to first plate the transformants in HL5 + 20 µg/mL uracil for 3 to 4 days without 5-FOA selection to allow the residual Pyr56 to decay. After this period, cells are titered, and new plates set up with 5×10^6 cells per dish in HL5 + 20 µg/mL ura + 100 µg/mL 5-FOA. It is a good idea to do a parallel control with no DNA added, so that the loss of parental cells can be monitored, and it is clear when ura⁻ cells are actually growing.

In a few days, the cells may aggregate because of the 5-FOA, preventing growth of normal cells. If they do, aspirate off the floating aggregates and medium, and replace with the same medium. The medium should be replaced every 4 to 7 days. Eventually, clones of cells will begin to grow in the dishes. These cells can be cloned and tested for uracil-dependent growth.

Reversion to Ura+

To revert ura⁻ cells, transform with the same pDU3B1 vector, and the next day, replace the medium with normal unsupplemented HL5. There is a significant delay in the loss of ura⁻ cells, so it is a good idea to do parallel control transformation, with no DNA, to know when ura⁺ cells are growing. The selection is more efficient if cells are plated in FM medium instead of HL5 because of the presence of some uracil even in unsupplemented in HL5.

Cloning Transformants

Clonal transformants can be obtained by several methods. Transformed cells generally form colonies on the petri dish. Cells from these colonies can be pulled from the plate with a pipettor and replated into a fresh dish. Alternatively, cells can be plated directly into 96-well microplates following transformation. A third method is to wait until transformed cells are stably growing, and then plate cells at low density on lawns of bacteria. However, cells are not under selection while growing on bacterial lawns. It seems that many surviving cells are not stably transformed, because if cloned too early on bacteria, most of the colonies will be drug sensitive when picked back to HL5 medium containing selective drug. It is also possible to isolate clones by plating cells in ultra-low gelling temperature agarose containing HL5 and selective drug. Colonies of cells will grow in the agar with minimal spreading (19). This technique has not been tested with nutritional selections.

To clone transformants in agarose, aliquots of 1.2% ultra-low gelling temperature agarose (see methods below) are melted in the microwave oven for about 15 to 30 seconds, and then cooled to 25°C. The agar will stay liquid at room temperature for several hours until spread on the plate. It seems that the increase in surface area-to-volume ratio of spreading on the plate greatly accelerates hardening of the agar. Antibiotics and then cells are added in a volume made up to 1 mL to bring the agar concentration to 1%, and then the medium is spread on the base agar. Normally, 3×10^5 to 1×10^6 cells are spread on each 100-mm base agar petri dish for transformation selections. Higher numbers seem to inhibit growth of the transformed colonies. A leveling plate in the cold room is used to ensure even plating, and the agar is allowed to harden for 15 to 30 minutes. The plates are incubated at 22°C in a humid chamber to prevent dehydration of the soft agar.

Colonies can be found with the microscope after about 5 days and by eye within 1 to 2 weeks. They come up faster with extrachromosomal vectors than with integrating vectors. The colonies can be picked by stabbing with a toothpick or wire needle and inoculating into a petri dish containing HL5 medium and drug used for selection. The colonies in the agar will eventually initiate development, but the process is asynchronous from colony to colony. For soft agar, use 30 µg/mL G418 or 100 µg/mL hygromycin.

STOCK SOLUTIONS

HL5 Growth Medium

5 g Proteose Peptone #2 (Difco Laboratories, Detroit, MI, USA)
5 g BBL Thiotone E
5 g Yeast extract
10 g Glucose
0.35 g Na_2HPO_4
0.35 g KH_22PO_4
Add distilled water to 1 L and adjust to pH 6.7 with HCl or NaOH.
Autoclave.
For uracil-dependent cells, add 20 µg/mL uracil before autoclaving.
For thymidine-dependent cells, add 40 µg/mL thymidine before autoclaving.

Cells can be grown in shaking suspension in HL5 medium or on the surface of either bacteriological- or tissue culture-treated plastic petri dishes. They will grow from a single cell up to a titer of approximately 3×10^6 cells/mL on a surface (10 mL of medium in a 100-mm dish) or 2×10^7 cells/mL in shaking suspension (25 mL in a 125-mL flask). Some parental strains form tight colonies on plastic surfaces, while others move rapidly over the surface of the dish so no colonies are observed.

Base Agar Plates for Soft Agar Cloning

7 g Proteose Peptone #2
7 g BBL Thiotone E
15.4 g Glucose
7.15 g Yeast extract
0.5 g Na_2HPO_4
0.48 g KH_2PO_4
8 g Bacto agar
Add distilled water to 1 L and adjust to pH 6.7 with HCl or NaOH.
Autoclave.
It is convenient to keep a bottle of medium plus agar that has been auto-

claved in the cold room. When plates are needed, melt the agarose in the microwave oven, allow it to cool, and then remove the needed amount to a sterile flask, add appropriate antibiotics, and pour the plates. Use 20 mL/petri dish containing 100 μg/mL dihydrostreptomycin sulfate and 50 μg/mL ampicillin (made up as a 100× stock and stored frozen). For selection in transformation, include 30 μg/mL G418. Plates can be poured the day before use (i.e., the same day as the transformation), but also can be made up in batches and stored in the cold for later use. We have not yet tested this procedure with other selection systems.

Soft Agar for Cloning

HL5 growth medium containing 1.2% ultra-low gelling temperature agarose (available from Fisher Scientific [Pittsburgh, PA, USA] or FMC Bio-Products [Rockland, ME, USA]).

Autoclave the agar plus medium in a bottle with a stir bar and then stir while hot to make sure that the agar is completely dispersed (it tends to clump at the bottom even after autoclaving). When dissolved, aliquot 5 mL/tube in sterile test tubes and store cold.

FM Growth Medium

This is a chemically defined growth medium in which axenic strains of *Dictyostelium* amoeba will grow. The major limitation is that, unlike HL5, cells will not grow clonally in this medium from single cells. They require an inoculum of more than about 1×10^4 cells/mL. The preparation of FM medium is described in detail on the *Dictyostelium* Web site (**http://dicty. cmb.nwu.edu/dicty/dicty.html**).

Electroporation

H-50 buffer, pH 7.0

20 mM 4-(2-hydroxyethyl)-1-piperazineethanesulfonic acid (HEPES)
50 mM KCl
10 mM NaCl
1 mM $MgSO_4$
5 mM $NaHCO_3$
1 mM NaH_2PO_4
Autoclave and store frozen or at 4°C.

CaPO₄ Coprecipitation

Bis-Tris HL5

2.1 g Bis-Tris (10 mM)
5 g Proteose Peptone #2
5 g BBL Thiotone E
5 g Yeast extract
10 g Glucose
   Add distilled water to 1 L and adjust to pH 7.25 with HCl or NaOH. After autoclaving, the pH will usually drop to 7.1. Check a small amount to be sure. If it is not, adjust the pH accordingly and then filter-sterilize. It is important that this solution be near pH 7.1, because the CaPO₄ precipitate is pH sensitive even after it is formed.

2× HBS

4.0 g NaCl (0.274 M)
0.18 g KCl (10 mM)
0.05 g Na₂HPO₄ (anhydrous) (1.4 mM)
2.5 g HEPES (42 mM)
0.5 g Glucose (12 mM)
Add distilled water to 250 mL and adjust pH to 7.1 ± 0.05 with NaOH.
Filter-sterilize, aliquot, and store frozen.

1.25 M CaCl₂

1.84 g CaCl₂ (dihydrate)
Add distilled water to 10 mL.
Filter-sterilize and store frozen.

30% Glycerol

30 mL Glycerol (Cat. No. G7893; Sigma, St. Louis, MO, USA)
70 mL Distilled water
Autoclave and store frozen.

Antibiotics

1000× G418

100 mg G418 (Geneticin from Sigma)
10 mL of 10 mM HEPES, pH 7.2
Filter-sterilize and store at 4°C (stable for months).

Table 1. *Dictyostelium* Transformation Vectors

Plasmid Name	Promoter	Tag	Selectable Marker	Size (kbp)	Extra Chromosomal	Cloning Sites	References
Transformation Vectors							
pTIKL			G418	10	Dpd1 Ori	S, Xb, Sp, B, Sm, R/H	41
pA15XPT1			G418	6	no	Bg, Pv	35
pDE102			Hyg	5	no	Sh, P, Sa, Xo/3, Sm, K, R	14
pDE109			Hyg	11	Dpd2 Ori	H, Sl, Xb/B, Sn, K, S, R	14
pBsr2 (pUC-Bsr)			Bsr	4.1	no	R, (Sc), K, Sm, (B)/H	40
pBsr3 (pUC-Bsr Bam)			Bsr	4.1	no	R, (Sc), K, Sm, B /H	1
pBsr BglII			Bsr	4.6	no	R, K, Sm, B, Xb/H	1
pDH29/pRHI13	ura		ura	5.7	no		R. Insall[a]
pRHI30	ura		ura	4.9	no		R. Insall[a]
pGTF2		promoter trap	Bsr	4.9	no	B, R	A. Kuspa[b]
pGEM25	thy		thy	6.7	no		13
Expression Vectors							
pDXA-3C	Act15	myc (C)	G418	6.1	with pREP	K, S, B, Xo, Ns, Xb	24
pDXA-3H	Act15	6 His (C)	G418	6.1	with pREP	K, S, B, Xo, Ns, Xb	24
pDXA-HC	Act15	His(N),myc(C)	G418	6.1	with pREP	K, S, B, Xo, Ns, Xb	24
pDXA-HY	Act15	His(N), YL1/2(C)	G418	6.1	with pREP	K, S, B, Xo, Ns, Xb	24
pDXD-3C	discoidin	myc (C)	G418	6.8	with pREP	K, S, B, Xo, Ns, Xb	24
pDXD-3H	discoidin	6His (C)	G418	6.8	with pREP	K, S, B, Xo, Ns, Xb	24
pDneoII	Act6	no	G418	6.2	no	S (2), Xb	41
pB18	Act15		G418	5.7	no		15

Table 1. (Continued)

Plasmid Name	Promoter	Tag	Selectable Marker	Size (kbp)	Extra Chromosomal	Cloning Sites	References
Expression Vectors (cont.)							
pAD-80-HA	Act15	HA (N)	G418	6.2	no	K, Bg, H, K	3
pVEIIc	discoidin	no	G418	6.1	no	K, S, B, Xb	2
pVEII ATG	discoidin	no	G418	6.1	no	K, Sc	34
pVEII ATG MCS	discoidin	no	G418	6.1	no	K, S, RV, R, B, Xb, Sc	37
pJK1	Act15	no	G418	12	yes		32
GFP Fusion Vectors							
pDNeoGFP	Act15	GFP (N)	G418	6.6	with pREP	Bg, Xo (Xb)	30
pDHygGFP	Act15	GFP (N)	Hyg	6.7	no	Bg	30
pDBsrGFP	Act15	GFP (N)	Bsr	6.4	no	Bg	30
p171/HygDGFP	discoidin	GFP	Hyg	7.2	no	B (N), S (C-term of GFP)	MacWilliams[b]
p172/NeoDGFP	discoidin	GFP	G418	6.9	no	Xb, B (N), S (C)	MacWilliams
p6OC2	Act15	GFP (N)	Bsr	11	Dpd1 Ori	Xo, Sm, K, Bg	21
discVsGFP	discoidin	GFP	G418	8.5	no	(Xo;C)/ Xb	11
discUbiGFP	discoidin	GFP	G418	8.6	no	(Xo;C)/ Xb	11
Tissue Specific Vectors							
pecmA/lGFP	ecmA	GFP (C)	G418	8.9	no	H, R (2), Bg, B / B, S, Xo	39
rasD-s65tGFP	rasD	GFP	G418		no		11
psA-GFP	psA	GFP	G418		no		11

Restriction enzymes: B, *Bam*HI; Bg, *Bgl*II; H, *Hind*III; K, *Kpn*I; Ns, *Nsi*I; P, *Pst*I; Pv, *Pvu*II; R, *Eco*RI; RV, *Eco*RV; S, *Sal*I; Sc, *Sac*I; Sh, *Sph*I; Sm, *Sma*I; Sp, *Spe*I.

(N) or (C) describes where the tag is relative to the restriction site for fusion.

[a]Insall, R.H., J. Borleis, and P.N. Devreotes. 1996. The *aimless* RasGEF is required for processing of chemotactic signals through G-protein-coupled receptors in Dictyostelium. Curr. Biol. *6*:719-729.

[b]Unpublished.

1000× Hygromycin B

250 mg Hygromycin B (Calbiochem-Novabiochem, San Diego, CA, USA)
10 mL Distilled water
Filter-sterilize and store in frozen 1-mL aliquots.
Thawed aliquots are stored at 4°C.

1000× Blasticidin

100 mg Blasticidin S (Calbiochem-Novabiochem)
10 mL water
Filter-sterilize and store frozen in 1-mL aliquots.
Thawed aliquots are stored at 4°C

1000x 5-Fluoro-orotic acid

Make up 100 mg in 1 mL dimethyl sulfoxide (DMSO) and store frozen.

STRAINS AND VECTORS

Most of the strains and vectors mentioned in this article are freely distributed within the *Dictyostelium* community (Table 1). Many are available from the author upon request (**knecht@uconnvm.uconn.edu**). Others can be obtained from the authors referenced in the bibliography. Phone numbers and e-mail addresses of *Dictyostelium* researchers can be found on the Dicty Web site (**http://dicty.cmb.nwu.edu/dicty/dicty.html**).

REFERENCES

1.**Adachi, H., T. Hasebe, K. Yoshinaga, T. Ohta, and K. Sutoh.** 1994. Isolation of *Dictyostelium discoideum* cytokinesis mutants by restriction enzyme-mediated integration of the blasticidin S resistance marker. Biochem. Biophys. Res. Commun. *205*:1808-1814.
2.**Blusch, J., P. Morandini, and W. Nellen.** 1992. Transcriptional regulation by folate-inducible gene expression in *Dictyostelium* transformants during growth and early development. Nucleic Acids Res. *20*:6235-6238.
3.**Burns, C.G., M. Reedy, J. Heuser, and A. De Lozanne.** 1995. Expression of light meromyosin in *Dictyostelium* blocks normal myosin II function. J. Cell Biol. *130*:605-612.
4.**Chang, A.C.M., R.M. Hall, and K.L. Williams.** 1991. Bleomycin resistance as a selectable marker for transformation of the eukaryote, *Dictyostelium discoideum*. Gene *107*:165-170.
5.**Chang, A.C.M., K.L. Williams, J.G. Williams, and A. Ceccarelli.** 1989. Complementation of a *Dictyostelium discoideum* thymidylate synthase mutation with the mouse gene provides a new selectable marker for transformation. Nucleic Acids Res. *17*:3655-3661.
6.**Colbere-Garapin, F., F. Horodniceanu, P. Kourilsky, and A.C. Garapin.** 1981. A new dominant hybrid selective marker for higher eukaryotic cells. J. Mol. Biol. *150*:1-14.
7.**Cole, R.A. and K.L. Williams.** 1988. Insertion of transformation vector DNA into different chro-

mosomal sites of *Dictyostelium discoideum* as determined by pulse field electrophoresis. Nucleic Acids Res. *16*:4891-4903.

8. Cox, D., J. Condeelis, D. Wessels, D. Soll, H. Kern, and D.A. Knecht. 1992. Targeted disruption of the ABP-120 gene leads to cells with altered motility. J. Cell Biol. *116*:943-955.

9. Coyne, C., U. Hamker, and H.K. MacWilliams. 1999. Green fluorescent proteins with short half-lives as reporters in *Dictyostelium discoideum*. Dev. Genes Evol. *209*:63-68.

10. De Lozanne, A. and J.A. Spudich. 1987. Disruption of the *Dictyostelium* myosin heavy chain gene by homologous recombination. Science *236*:1086-1091.

11. Deichsel, H., S. Friedel, A. Detterbeck, C. Coyne, U. Hamker, and H.K. MacWilliams. 1999. Green fluorescent proteins with short half-lives as reporters in *Dictyostelium discoideum*. Dev. Genes Evol. *209*:63-68.

12. Dingermann, T., N. Reindl, T. Brechner, H. Werner, and K. Nerke. 1990. Nonsense suppression in *Dictyostelium discoideum*. Dev. Genet. *11*:410-417.

13. Dynes, J.L. and R.A. Firtel. 1989. Molecular complementation of a genetic marker in *Dictyostelium* using a genomic DNA library. Proc. Natl. Acad. Sci. USA *86*:7966-7970.

14. Egelhoff, T.T., S.S. Brown, D.J. Manstein, and J.A. Spudich. 1989. Hygromycin resistance as a selectable marker in *Dictyostelium discoideum*. Mol. Cell. Biol. *9*:1965-1968.

15. Johnson, R.L., R.A. Vaughan, M.J. Caterina, P.J.M. van Haastert, and P.N. Devreotes. 1991. Over-expression of the cAMP receptor 1 in growing *Dictyostelium* cells. Biochemistry *30*:6982-6986.

16. Kalpaxis, D., I. Zundorf, H. Werner, N. Reindl, E. Boy-Marcotte, M. Jacquet and T. Dingermann. 1991. Positive selection for *Dictyostelium discoideum* mutants lacking UMP synthase activity based on resistance to 5-fluoroorotic acid. Mol. Gen. Genet. *225*:492-500.

17. Katz, K.S. and D.I. Ratner. 1988. Homologous recombination and the repair of double-strand breaks during cotransformation of *Dictyostelium discoideum*. Mol. Cell. Biol. *8*:2779-2786.

18. Knecht, D.A., S.M. Cohen, W.F. Loomis, and H.F. Lodish. 1986. Developmental regulation of *Dictyostelium discoideum* actin gene fusions carried on low-copy and high-copy transformation vectors. Mol. Cell. Biol. *6*:3973-3983.

19. Knecht, D.A., J.H. Jung, and L. Matthews. 1990. Quantification of transformation efficiency using a new method for clonal growth and selection of axenic *Dictyostelium* cells. Dev. Genet. *11*:403-409.

20. Kuspa, A. and W.F. Loomis. 1992. Tagging developmental genes in *Dictyostelium* by restriction enzyme-mediated integration of plasmid DNA. Proc. Natl. Acad. Sci. USA *89*:8803-8807.

21. Larochelle, D.A., K.K. Vithalani, and A. De Lozanne. 1997. Role of *Dictyostelium* racE in cytokinesis: mutational analysis and localization studies by use of green fluorescent protein. Mol. Biol. Cell. *8*:935-944.

22. Loomis, W.F. 1996. Genetic networks that regulate development in *Dictyostelium* cells. Microbiol. Rev. *60*:135.

23. Loomis, W.F. 1998. The *Dictyostelium* genome sequencing project. Protist *149*:209-212.

24. Manstein, D.J., H.P. Schuster, P. Morandini, and D.M. Hunt. 1995. Cloning vectors for the production of proteins in *Dictyostelium discoideum*. Gene *162*:129-134.

25. Metz, B.A., T.E. Ward, D.L. Welker, and K.L. Williams. 1983. Identification of an endogenous plasmid in *Dictyostelium discoideum*. EMBO J. *2*:515-519.

26. Morrison, A., R. Marschalek, T. Dingermann, and A.J. Harwood. 1997. A novel, negative selectable marker for gene disruption in *Dictyostelium*. Gene *202*:171-176.

27. Nellen, W. and R.A. Firtel. 1985. High-copy-number transformants and co-transformation in *Dictyostelium*. Gene *39*:155-163.

28. Nellen, W., C. Silan, and R.A. Firtel. 1984. DNA-mediated transformation in *Dictyostelium discoideum*: regulated expression of an actin gene fusion. Mol. Cell. Biol. *4*:2890-2898.

29. Noegel, A., D.L. Welker, B.A. Metz, and K.L. Williams. 1985. Presence of nuclear associated plasmids in the lower eukaryote *Dictyostelium discoideum*. J. Mol. Biol. *185*:447-450.

30. Pang, K.M., M.A. Lynes, and D.A. Knecht. 1999. Variables controlling the expression level of exogenous genes in *Dictyostelium*. Plasmid *41*:187-197.

31. Pellicer, A., M. Wigler, R. Axel, and S. Silverstein. 1978. The transfer and stable integration of the HSV thymidine kinase gene into mouse cells. Cell *14*:133-141.

32. Pitt, G.S., N. Milona, J. Borleis, K.C. Lin, R.R. Reed, and P.N. Devreotes. 1992. Structurally distinct and stage-specific adenylyl cyclase genes play different roles in *Dictyostelium* development. Cell *69*:305-315.

33. Raper, K.B. 1940. Pseudoplasmodium formation and organization in *Dictyostelium discoideum*. J. Elisha Mitchell Sci. Soc. *56*:241-282.

34. Rebstein, P.J., G. Weeks, and G.B. Spiegelman. 1993. Altered morphology of vegetative amoeba induced by increased expression of the *Dictyostelium discoideum* ras-related gene rap1. Dev. Genet. *14*:347-355.

35. Scherczinger, C.A., A.A. Yates, and D.A. Knecht. 1992. Variables affecting antisense RNA inhibition of gene expression. Ann. NY Acad. Sci. *660*:45-56.

36. Schlatterer, C., G. Knoll, and D. Malchow. 1992. Intracellular calcium during chemotaxis of *Dictyostelium discoideum*—a new fura-2 derivative avoids sequestration of the indicator and allows long-term calcium measurements. Eur. J. Cell Biol. *58*:172-181.

37. Seastone, D.J., E. Lee, J. Bush, D. Knecht, and J. Cardelli. 1998. Overexpression of a novel Rho family GTPase, RacC, induces unusual actin-based structures and positively affects phagocytosis in *Dictyostelium discoideum*. Mol. Biol. Cell. *9*:2891-2904.

38. Shah-Mahoney, N., T. Hampton, R. Vidaver, and D. Ratner. 1997. Blocking the ends of transforming DNA enhances gene targeting in *Dictyostelium*. Gene *203*:33-41.

39. Stege, J.T., G. Shaulsky, and W.F. Loomis. 1997. Sorting of the initial cell types in *Dictyostelium* is dependent on the tipA gene. Dev. Biol. *185*:34-41.

40. Sutoh, K. 1993. A transformation vector for *Dictyostelium discoideum* with a new selectable marker Bsr. Plasmid *30*:150-154.

41. Uyeda, T.Q.P., P.D. Abramson, and J.A. Spudich. 1996. The neck region of the myosin motor domain acts as a lever arm to generate movement. Proc. Natl. Acad. Sci. USA *93*:4459-4464.

42. Witke, W., W. Nellen, and A. Noegel. 1987. Homologous recombination in the *Dictyostelium* alpha-actinin gene leads to an altered mRNA and lack of the protein. EMBO J. *6*:4143-4148.

5 | Introduction of DNA into Cultured Mammalian Cells

Pamela A. Norton

Division of Gastroenterology and Hepatology, Department of Medicine, Thomas Jefferson University, Philadelphia, PA, USA

INTRODUCTION

The last two decades have seen the cloning and manipulation of DNA go from a complex and laborious undertaking to a fairly routine set of procedures. Many cloning projects have been motivated by the goal of transferring the cloned DNA sequences back into intact cultured cells of metazoan origin in order to assess function. Thus, the development of methods for the introduction of DNA into cultured cells, especially mammalian cells, has proceeded in parallel with advances in molecular cloning techniques. The process of introducing DNA into vertebrate cells is generally referred to as transfection, although some authors have referred to the same process as transformation, by analogy with DNA transfer in prokaryotes. In this chapter, the term transfection will be used exclusively to avoid confusion, as transformation can have a distinct meaning with regard to mammalian cells relating to growth and morphologic state (18).

Ideally, DNA will be transferred to cells with 100% efficiency (all target cells will acquire and express the introduced DNA) and with minimal toxicity (all cells will survive the procedure). A number of methods to introduce DNA directly into mammalian cells have been described, but virtually all fall short of these ideals. Indeed, increased transfection efficiency often correlates with increased toxicity, necessitating a tradeoff. Despite continual refinements, several different transfection methods remain in common use, as the optimal procedure for one cell type is often unsuited to another. In addition, certain applications might result in a preference for specific procedures. This chapter reviews the most commonly encountered methods, with considera-

Gene Transfer Methods
Edited by Pamela A. Norton and Laura F. Steel
©2000 Eaton Publishing, Natick, MA

tion of their relative merits both in principal and in practice. In addition, the mechanisms of DNA uptake are discussed, as further research in this area will likely lead to improved methods in the future. Finally, in addition to presenting various transfection protocols, procedures are included for assessing transfection efficiency and the generation of stable lines of transfected cells.

OVERVIEW AND HISTORY OF TRANSFECTION METHODS

General Considerations and Choosing a Method

Transfection methods can be grouped simplistically into two categories: physically mediated delivery or chemically mediated delivery. Physical methods include electroporation, particle bombardment (gene guns), and direct microinjection; all permit DNA entry into the cell by temporarily breaching the integrity of the plasma membrane. Chemically mediated facilitation of DNA uptake includes agents such as inorganic calcium phosphate, positively charged polymers such as DEAE dextran, and mixtures of lipids, often referred to as liposomes. Despite the diversity of these agents, the mechanics of their use tend to be more similar than different. The target cells are usually passaged one day prior to transfection, so that they are actively growing when exposed to DNA. Typically, DNA and the facilitating agent are mixed at room temperature briefly (5–60 min) and then applied to cells. The main procedural differences arise with the latter step, depending on whether the cells are adherent or grow in suspension. The cells are incubated in the presence of the DNA for a period of 1 to 24 hours, at which point they are transferred to their preferred growth medium. Incubation proceeds for one or more additional days, depending on the nature of the experiment.

Selection of the optimal transfection method remains an empirical process, as the efficiency of a procedure tends to be highly dependent on the cell type to be used. When planning to transfect a given cell type for the first time, reference to the literature and/or experienced colleagues may prove valuable, especially for commonly used cell types. However, it should be understood that the same cell line might behave quite differently from one laboratory to another, because of differences in routine culture conditions, and this might necessitate modifications of procedures established in other laboratories. Nevertheless, the reader should be reassured that most cell lines can be transfected successfully, if not optimally, with most of the methods described. In the absence of any prior information regarding the optimal transfection method for a given cell type, or if the intended target is a primary cell type, an initial comparison of several procedures may be warranted. In recognition of this situation, some commercial vendors market sampler kits designed to allow head-to-head comparisons of different reagents (e.g., Reference 25). Purchasing such a kit is an attractive option if a laboratory has no prior experience with any of the transfection methods described. In addition, the possibility for obtaining trial sizes or free samples of various reagents should be explored.

Other considerations to weigh include cost and convenience; many commercial reagents and kits are very convenient but can prove expensive if the number of experiments planned is large. Certain procedures (e.g., electroporation) require a specialized apparatus, and accessibility might prove limiting. Another factor is that many cells do poorly in the absence of serum for any length of time, whereas a number of the commercially available lipid reagents are most effective in the absence of serum. Reagents that tolerate the presence of serum during the initial exposure of cells to DNA might be preferred for delicate cells. The more information and experience a research team has with handling and culturing the cell type of interest, the easier it will be to choose an appropriate transfection method and the greater the possibility of success. A description of the origin and usage considerations for each of the major methods follows.

Polycation-Mediated DNA Uptake

The earliest report of polycation-mediated DNA uptake employed the derivatized polysaccharide diethylaminoethyl dextran (DEAE dextran). Extrapolating from the observation that polycations stimulate infectivity of nucleic acid isolated from RNA viruses, McCutchan and Pagano (39) demonstrated that DEAE dextran facilitated the uptake of viral DNA by mammalian cells. Efficiency was dependent on both the concentration and the chain length of the dextran. It was demonstrated subsequently that brief treatment of cells with either dimethyl sulfoxide (DMSO) or glycerol after exposure to DNA further enhanced transfection efficiency (35,60). Other polycations, such as protamine or polyethylenimine (PEI), a branched polymer, also can be used to introduce DNA into cells. PEI, and probably the other polycations, appears to both condense DNA via ionic interactions and promote interaction with the negatively charged cell surface (3). Following endocytic uptake (see the section entitled Mechanisms of DNA Uptake), the excess of protonatable nitrogens should serve to buffer the lysosomal compartments, decreasing DNA degradation. Although the use of these agents is similar to DEAE dextran (DNA is combined with the polyamine under physiological salt concentrations), the optimal ratio of polyamine to DNA will vary, in part due to the precise charge ratio of the cationic agent to the negatively charged phosphates of the nucleic acid. Some reports suggest that the combination of polycations with lipids may improve delivery when compared with either substance alone (see the section entitled Lipid-Mediated DNA Uptake). However, DEAE dextran and PEI are both inexpensive, which is a factor in considering their usage.

Polyamidoamine dendrimers are complex polycations consisting of branches with terminal charged amino groups radiating from a central core, forming a rough sphere (26,29). The DNA is compacted within the dendrimer structure, and the net positive charge is thought to both facilitate association with the plasma membrane as well as to buffer lysosomal pH. These reagents are used in essentially the same manner as the polyamine reagents, and similar principles should apply. Although de novo synthesis of these reagents is complex, activated dendrimer reagents are available commercially (Table 1), and this category of transfection reagent is likely to continue to evolve.

Calcium Phosphate-Mediated DNA Uptake

Based on the requirement for divalent cations in bacterial transformation, Graham and Van Der Eb (24) reasoned that a similar strategy might enhance the uptake of DNA by mammalian cells. In a detailed study, this prediction was proved to be correct, and a transfection protocol was formulated that since has been followed widely (24). Briefly, the DNA is diluted into a solution buffered within a narrow pH range and containing a low concentration of phosphate ions. Calcium ions, in the form of $CaCl_2$, are then added, and the DNA is coprecipitated with the inorganic salts. The precipitates are applied to cells and are apparently endocytosed during subsequent incubation. Further enhancement of transfection efficiency can be obtained by brief treatment of the cells with DMSO after removal of nonadsorbed DNA complexes (59). A more recent version of this procedure using BES (*N,N*-bis[2-hydroxyethyl]-2-aminoethanesulfonic acid) as the buffer in place of HEPES (N-[2-hydroxyethyl]piperazine-*N'*-[2-ethanesulfonic acid]) may be more efficient under certain circumstances (6).

Many workers have favored $CaPO_4$–DNA precipitation for the production of stably transfected cells, as coprecipitation of multiple molecules occurs with high frequency, permitting the cotransfection of a selectable marker along with the gene of interest (48,65). The reagents are inexpensive, but can require some effort to prepare, because of the need for precise control of pH. In addition, this method tends to require a large mass of DNA to achieve efficient precipitate formation. This results in the use of large amounts of the plasmid DNA of interest or the inclusion of carrier DNA in the mixture. Finally, some cell types appear to be completely refractory to $CaPO_4$-mediated transfection. This variability has resulted in a gradual favoring of lipid-based transfection methods, but $CaPO_4$–DNA precipitation remains in use in many laboratories.

Lipid-Mediated DNA Uptake

The earliest described use of lipids for nucleic acid delivery was phosphatidylserine-mediated delivery of poliovirus RNA (66). Although this and other early studies were motivated by a desire to use nucleic acids as a sensitive measure of the intracellular delivery of liposome contents, it was soon recognized that gene delivery was an important goal in of itself. Many lipid preparations were tested in an effort to improve the efficiency of delivery (17). An important development was the inclusion of a cationic lipid, which appears to fulfill two functions: *(i)* neutralizing the net positive charge of the DNA, and *(ii)* complexing with the positively charged cell surface (13,15). In the initial report, the cationic lipid N-(1-(2,3-dioleyloxy)propyl)-N,N,N-trimethylammonium chloride (DOTMA) was mixed in a 1:1 ratio with the neutral lipid L-dioleoyl phosphatidylethanolamine (DOPE or PtdEtn).

The preparation of lipoplexes, a term proposed for the complex between a cationic lipid and a nucleic acid (12), is straightforward in principal: typically, DNA and lipid are diluted separately, then mixed before application to cells. However, preparation of the lipid component presented a barrier to widespread adoption of lipid-mediated DNA transfection. Although DOTMA and other lipids may be synthesized in the laboratory (13), the procedures involved may be unfamiliar to a laboratory with expertise in cellular biology and interest in performing gene transfer experiments. Fortunately, a number of reagents have become available commercially in recent years, resulting in a dramatic increase in the popularity of lipid-mediated gene transfer methods over the last ten years. The listing of commercial reagents compiled in Table 1 may not be comprehensive, as new products appear with high frequency. The use of commercially prepared reagents offers both convenience and reproducibility, although these advantages come at some expense. Cost will not be prohibitive unless very large quantities are needed, as when the ultimate experimental endpoint will include in vivo gene delivery. In the latter event, a recent report suggests an inexpensive method for the preparation of large amounts of lipoplexes using inexpensive disposable syringes and a plastic T-tube as a mixer (27).

If a commercial reagent is chosen, then the manufacturer's recommended procedures should be followed as a starting point in determining the ratio of DNA to lipid. One critical variable is whether the presence of serum in the medium will prove detrimental to gene delivery. Many cells tolerate the removal of serum for the few hours needed for the transfection procedure, but if the intended target cell does not fall into this category, a reagent that will perform in the presence of serum should be chosen. In addition, serum-tolerant liposomes will not require supplementation or replacement of medium to restore the serum needed for growth, further simplifying the experimental procedure.

There is some evidence that polyamines and liposomes can cooperate in the introduction of DNA into cells. High molecular weight cations appear to facilitate DNA condensation, reducing particle size of lipoplexes (21). Similarly, encapsulating PEI–DNA complexes into liposomes can augment

Table 1. Suppliers of Commercial Liposomal Reagents for Transfection of Mammalian Cells

Supplier	Product	Type
Bio-Rad Laboratories (Hercules, CA, USA)	Cytofectene™	Proprietary cationic lipid:DOPE, 1:1
Roche Molecular Biochemicals (Indianapolis, IN, USA)	DOSPER DOTAP FuGENE™6	Polycationic liposome Monocationic liposome Proprietary lipids
Clontech Laboratories (Palo Alto, CA, USA)	CLONfectin™	Cationic amphiphilic lipid
Gene Therapy Systems (San Diego, CA, USA)	GenePORTER™ GenePORTER™ 2	Proprietary lipid (Direct hydrophilic conjugation) DOPE:proprietary cationic lipid
Glen Research (Sterling, VA, USA)	Cytofectin GS/GSV	Cationic lipid:DOPE, 2:1 (GSV, vesicle form)
Invitrogen (Carlsbad, CA, USA)	PerFect™ Lipids: pFx™-1 to pFx™-8	Various combinations of cationic lipids and/or DOPE
JBL Scientific (San Luis Obispo, CA, USA)	Eu·Fectin™ 1-11	Various proprietary lipids of varying net charge, shape, and size
LifeTechnologies (Rockville, MD, USA)	CellFECTIN™ DMRIE-C LIPOFECTACE™ LIPOFECTAMINE LIPOFECTIN®	Cationic lipid TM-TPS:DOPE, 1:1.5 Cationic lipidDMRIE:cholesterol, 1:1 Cationic lipid DDAB:DOPE, 1:2.5 Cationic lipid DOSPA:DOPE, 3:1 Cationic lipid DOTMA:DOPE, 1:1
PanVera (Madison, WI, USA)	TransIT-100	Cationic lipid
Promega (Madison, WI, USA)	Tfx™-10 Tfx™-20 Tfx™-50 TransFast™ Transfectam	Cationic lipid combined with varying amounts of DOPE Cationic lipid:DOPE Lipopolyamine DOGS
Qiagen (Valencia, CA, USA)	Effectene™	Proprietary lipid-based reagent
Sigma (St. Louis, MO, USA)	DOTAP DOTAP-chloride Escort™	Monocationic liposome Monocationic liposome, salt form Liposome
Stratagene (La Jolla, CA, USA)	LipoTaxi™	Liposome

transfection efficiency (2), suggesting that the combination of the polyamines and the lipids offers benefits that surpass either agent alone. Lipopolyamines are a hybrid between the two classes of molecules, as their name implies. One such compound, dioctadecylamidoglycylspermine, or DOGS, consists of a spermine head attached to a lipid tail (34). A complete description of the synthesis of DOGS is available (34), although as with the cationic lipids, the procedure may prove formidable for many laboratories. Alternatively, the compound is available commercially as Transfectam® (Promega, Madison, WI, USA) (Table 1). In theory, the lipopolyamines combine the ability of the polyamine moiety to compact DNA and neutralize its negative charge with the liposomal structures. The procedures offered below for PEI- or lipid-mediated transfection should be adaptable to use with DOGS as well.

Electroporation

Cells exposed to high-voltage electric fields will take up exogenous DNA in a process referred to as electroporation (45). The electric field induces the formation of transient breaks, or pores, in the plasma membrane, allowing DNA and/or other molecules to pass into the cytoplasm. This relatively non-specific mechanism of action may account for its applicability to a wide variety of cell types (1,51,52).

At present, a number of electroporation devices are commercially available; those that deliver a pulse that decays exponentially by discharging a capacitor are less expensive and more widely used, but devices that deliver a rectangular or square wave pulse also are available. For mammalian cell transfection, the cells and DNA are combined in a small volume in a chamber such as a cuvette with a 0.4-cm gap between the electrodes. The charge is then applied, and after a brief incubation to allow the DNA to cross the membrane, the cells are restored to normal culture conditions. Thus, the procedure is quite rapid, especially for cells that normally grow in suspension, such as lymphocytic cells, which can be difficult to transfect by other methods. On the other hand, treatment of many samples could become cumbersome. Another apparent advantage of electroporation over the chemical methods is that even larger DNA molecules such as cosmids appear to be delivered efficiently (1), probably because of the minimal handling involved. The main potential drawback to the method is the need for specialized equipment; note that not all electroporation devices are designed for use with mammalian cells.

Although electroporation is relatively simple in concept, it presents several considerations that are unique to this method. Electric field strength and pulse length are both key parameters that must be optimized to maximize transfection efficiency without excessive cell death. For the more commonly used exponential decay machines, the maximal initial voltage delivered is set, and the duration of the decay is the product of the capacitance setting and the resistance of the sample. Successful transfection of several mammalian cell lines has been performed in HEPES-buffered saline at approximately 500 V/cm and 500 to 1000 μF, yielding a time constant of approximately 7 milliseconds (7), but higher voltage, low capacitance settings (e.g., 8 kV/cm and

20 mF) also prove effective (51). Alteration of reaction volume also may affect resistance. Another decision is the choice of electroporation buffer; phosphate or HEPES-buffered salines are good high-salt buffers, but some cells are poorly viable in buffer alone and fare better in serum-free culture medium. As an aid to simplifying these choices, two instrument manufacturers have compiled successful electroporation protocols for numerous cell types; these can be found at (**www.bio-rad.com**) and (**www.btxonline.com**).

The other physical methods of microinjection and particle bombardment rarely are used to deliver DNA to cultured mammalian cells, and thus procedures are not offered in this chapter. Although the efficiency of DNA transfer by microinjection into the nucleus is high (4), the procedure is laborious and can only be used to deliver DNA to a limited number of cells. Microinjection commonly is used for embryos, however (see Chapters 6 and 7); the larger size of the target cell facilitates injection, and analysis of each gene transfer event on an individual basis is desirable. Similarly, particle bombardment has been used by some workers for delivery of DNA to plants (Chapter 8). If the equipment is available, interested readers are directed to these chapters as a starting point.

Viruses

Although generally not considered transfection methods, viral vector systems deserve brief mention as a means of gene delivery. Viruses are highly adapted to the process of nucleic acid transfer; although they generally transfer their own genetic material exclusively, some viral genomes have been adapted so as to transmit heterologous DNA. Viral vectors have the ability to transfer DNA to a high fraction of cells, generally with little toxicity. Common vectors used for gene transfer in cell culture are derived from retroviruses (40), but adenovirus and other agents have been explored for applications where the end use is in vivo delivery. For instance, introduction of genes into live chick embryos has benefited from the use of replication-competent retrovirus vectors, which can be grown to relatively high titers, allowing efficient delivery (41). Retroviruses have also been useful in developmental studies in other species such as zebrafish (20). One negative feature is that viral vectors are almost always hampered by limitations on the size of the insert that can be transferred. In addition, assembly of constructs is often more complicated. However, there are some instances where the advantages of high frequency delivery may outweigh the disadvantages, such as delivery of genes to humans both in vivo and ex vivo (Chapter 9). Thus, although viral vectors are well suited to many applications, direct transfer of DNA using the procedures described below is often satisfactory and almost always simpler.

Determining Transfection Efficiency

When evaluating different transfection methodologies to select the one most appropriate for a given cell type, it is desirable to have a simple method of assessing transfection efficiency. One commonly used method for adherent cells is to introduce a plasmid containing a *lacZ* reporter gene, as the product, β-galactosidase is readily detected by incubation of fixed cells with the colorimetric substrate X-gal (5-bromo-4-chloro-3-indolyl-β-D-galactoside). Transfected cells turn blue, whereas untransfected cells remain unstained, and simple inspection of several fields under the microscope allows a quantitative assessment of the fraction of cells that acquired and expressed the DNA. There are several commercially available plasmids in which *lacZ* has been placed under the control of a strong promoter. Alternatively, fluorescent β-galactosidase substrates such as fluorescein di-β-D-galactopyranoside are available that are suitable for detection by fluorescence-activated cell sorting (FACS) (16,46). Finally, transfection of the gene encoding green fluorescent protein can be used to assess transfection efficiency using FACS analysis or fluorescence microscopy (5). An advantage of the latter procedures is that cell viability can be retained.

MECHANISMS OF DNA UPTAKE

It is convenient to think about the chemical facilitation of gene transfer in terms of the various obstacles that the DNA must overcome to be successfully expressed in the target cell (Figure 1). Although the complexes formed by various cationic- and lipid-based reagents may be chemically distinct, certain features of DNA delivery appear to be common. First, the DNA is complexed with the chemical agent, which should compact the DNA, protect it from degradation, and facilitate interaction with the negatively charged cell surface. Next, the DNA must cross the plasma membrane and enter the cell; this appears to occur most frequently by endocytosis. Following internalization via endocytosis, the DNA must be released from the vesicles and avoid being degraded in lysosomes. In addition, the DNA must escape degradation in the cytoplasm. Finally, the DNA must enter the nucleus, either as naked DNA or in the form of a complex. Most measures of transfection efficiency do not distinguish which of these steps are potentially rate limiting. However, a review of what is known about DNA complex formation and transfection mechanisms offers some insight into these events.

The structure of DNA–lipid complexes has been investigated, and plausible models for such interactions have been discussed in detail (57). It has been proposed that a negatively charged plasmid DNA molecule can interact with the surface of one or more positively charged spherical liposomes (15). However, DNA may also become sandwiched between membrane bilayers (53). Studies with cationic lipids support the latter configuration for DNA:lipid charge ratios similar to those typically used for DNA transfection, although at lower lipid levels, less compact associations were observed (22). Indeed, the topology of the lipid–DNA complexes can vary dramatically depending on both the amounts and the composition of the lipids. DOPE tends to form a nonlamellar, two-dimensional hexagonal phase under physiological conditions (33). Mixtures of cationic DOTAP (dioleoyl trimethylammonium propane) and DOPE can condense DNA into a columnar hexagonal array, in which the cationic groups are within the interior of the structure in close association with the DNA (28). In contrast, the less effective neutral lipid DOPC (dioleoyl phosphatidylcholine) packages the DNA into a multilamellar structure in which the DNA is trapped between lipid bilayers. These phase transitions have important functional ramifications. DOPE, but not DOPC, can improve transfection efficiency when mixed with polycationic lipids, even if supplied as a separate helper liposome (11,14,72). Possible roles for DOPE in the transfection process are considered below.

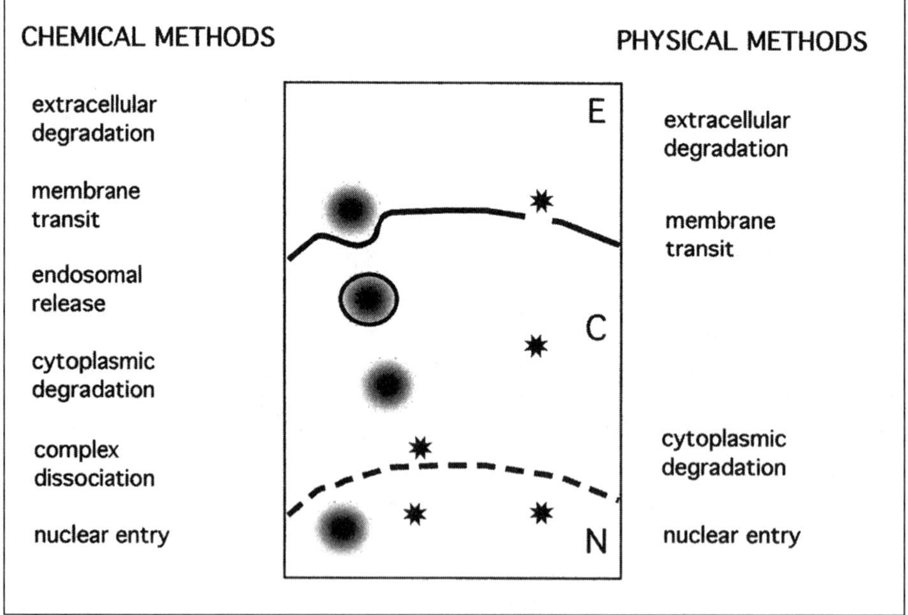

Figure 1. Barriers to successful transfection. The various stages that can be rate limiting for transfection are compared for chemical versus physical methods. Naked DNA is represented by the star, the plasma membrane by the solid line, and the nuclear envlope by the broken line. Note that for the chemical methods, there is uncertainty regarding the location at which the DNA becomes dissociated from the facilitating agent, represented by the shaded sphere. E, extracellular environment; C, cytoplasm; N, nucleus.

The mechanism by which calcium phosphate ($CaPO_4$) facilitates DNA uptake is understood in part. DNA is entrapped by coprecipitation with the $CaPO_4$, which renders it relatively resistant to nucleases such as those present in serum. The DNA is taken up by the cells in the form of the precipitate, apparently by endocytosis of the aggregates (36). Similarly, cationic polymers such as PEI interact with the DNA, condensing it, and reducing the electrostatic repulsion between the negatively charged DNA and the similarly charged cell surface, permitting endocytosis (23). Complexes of DNA with polyamidoamine dendrimers also appear to be taken up via endocytosis (29). The lipid agents, especially cationic lipids, reduce charge repulsion by compacting the DNA (22). Cationic liposomes bind to the cell surface, apparently via ionic interactions, and are taken up by the cells via nonreceptor-mediated endocytosis (14,19,31,70,72) or phagocytosis (38). Thus, although the complexes formed by various cationic and lipid reagents may be distinct, they all appear to promote uptake largely via endocytic processes.

Endocytosis of the various DNA complexes described above appears to be a relatively nonspecific process, although cell surface proteoglycans have been implicated in cationic liposome uptake (44). In an effort to increase the efficiency of DNA uptake, a number of strategies have been devised to take advantage of receptor-mediated endocytosis. For instance, the transferrin receptor has been targeted by complexing DNA with poly-L-lysine conjugated to transferrin, the ligand for this receptor (10). Similarly, the asialoglycoprotein receptor on hepatocytes facilitates the uptake of complexes of DNA and a desialylated protein linked to poly-L-lysine (67). However, DNA uptake does not appear to be the major barrier to optimal transfection efficiency (70), suggesting that the additional steps needed to prepare the ligand-containing complexes are unnecessary for most cells. Nevertheless, receptor-mediated targeting may prove valuable for efficient cell-type-specific delivery in vivo.

The fate of the DNA once inside the cell also must be considered. Cationic lipids are able to deliver DNA to as many as 100% of treated cells, but subsequent barriers prevent most of these cells from actually expressing the transgene (61). Similarly, most cells transfected using $CaPO_4$-mediated delivery acquire cytoplasmic DNA, suggesting that uptake is not limiting (36). However, it appears that only a small fraction of the intracellular DNA enters the nucleus, and much of the DNA taken up appears to remain in endosomal vesicles. It is possible that this vesicular DNA can be delivered to the nucleus via fusion with the nuclear envelope (19). Alternatively, DNA may need to be released from the endosome into the cytoplasm prior to nuclear import. It has been proposed that anionic lipids play a role in the release of DNA from endocytosed cationic liposomes by neutralizing the positive charge of the latter (68). However, several lines of evidence point to a role for DOPE in DNA release from endosomes following cationic lipid-mediated transfection. First, a peptide that inhibits the formation of the inverted hexagonal phase decreases transfection activity (72). Second, the neutral lipid DOPE appears to have endosomolytic activity, which may release the DNA into the cytoplasm (11). Finally, the cationic lipid- and DOPE-containing hexagonal complexes fuse readily in vitro, releasing the components of the complex (28). Thus, the transfection-enhancing properties of DOPE may be due to more efficient escape from the endosomal pathway. This may be a critical step for transfection efficiency, as DNA must either be released from the endosomal compartment prior to fusion with lysosomes, or the DNA must remain complexed in such a way as to prevent degradation; both events may occur to some extent.

Ultimately, DNA must reach the nucleus, whether free or as a complex. Microinjection of naked DNA into the cytoplasm results in a much lower level of transfection efficiency than injection directly into the nucleus (4,70). Another study suggests that degradation of naked DNA in the cytosol by resident nucleases represents a significant barrier to DNA delivery to the nucleus (30). Thus, retention of the DNA in a complex with the facilitating polycations or lipids might be advantageous, offering protection from degradation. In keeping with this idea, injection of PEI–DNA complexes into the cytoplasm resulted in higher levels of reporter gene expression than injection of naked DNA (50). The direct injection of PEI–DNA complexes into the nucleus demonstrated that the polycation did not interfere with gene expression. In contrast, intranuclear injection of DNA complexed with either the cationic lipid DOTAP, or the lipopolyamine DOGS interfered with gene expression (50,70). Thus, the intracellular pathways that DNA takes after endocytic internalization might not be identical for all transfection facilitators, and cell-type-specific differences in transfection efficiency might reflect, in part, differences in intracellular trafficking patterns and or kinetics.

Finally, it is interesting to note that most transfection protocols, including those presented in this chapter, suggest that cells be in an actively growing state. The breakdown of the nuclear envelope that occurs during mitosis might facilitate DNA entry into the nucleus (70). Recent studies also support a correlation between cell proliferation and increased gene expression (42 and references therein). As a means to improve DNA transport to the nucleus, peptides that promote nuclear localization have been incorporated into lipoplexes (71). Such procedural refinements will continue to improve transfection efficiencies.

In summary, many gaps remain in our understanding of the mechanisms whereby cells take up exogenous DNA. The above discussion has focused on the chemical transfection methods, as the barriers are more numerous and better studied. However, it is obvious that the physical methods are subject to a subset of the same limitations (Figure 1). The time of exposure of naked DNA outside of the cells is brief, limiting extracellular degradation. The physical transfection methods all involve temporarily breaching the cell membrane, bypassing the endosomal compartment, but cytoplasmic degradation may represent a significant source of DNA loss. In addition, DNA must enter the nucleus, and as discussed above, this can be a rate-limiting step. The interest in gene therapy applications will serve to propel further research into the processes involved in intracellular gene delivery, which should benefit all who are interested in using transfection methods.

APPLICATIONS

The possible experimental applications of transfection techniques are numerous, including the study of aspects of gene expression, DNA replication, and cell growth and differentiation. However, a certain fundamental distinction can be drawn between the two major categories of experimental outcomes: transient versus stable introduction of DNA. In most cases, one is generally concerned with achieving the highest transfection efficiency possible—that is, introducing the DNA of interest into the highest percentage of cells possible. Some combinations of transfection procedure and cell type can achieve efficiencies in excess of 50% when expression of the introduced DNA is measured 1 to 3 days posttransfection. The vast amount of this DNA will be lost over the course of several days. However, it is sometimes desirable to make stable transfectants, in which the DNA has been integrated into the host genome, so that all daughter cells inherit the foreign gene.

Stable transfection occurs with the integration of the exogenous DNA into the host cell DNA. This process is inefficient and needs to be accompanied by either the conferral of a novel phenotype on the cells by the introduced gene or, more generally, the simultaneous introduction of a gene that confers a selectable phenotype. Typically, selectable markers confer resistance to a pharmacologic agent, which is normally toxic to the cells. The most commonly used selectable marker is the aminoglycoside phosphotransferase (*neo*) gene, which confers resistance to the drug geneticin/G418 (58). Another gene that is useful in mammalian cells is hygromycin-B-phosphotransferase (*HPH*), which confers resistance to hygromycin-B (47). One to three days posttransfection, the appropriate drug is added to the culture medium. Nontransfected cells will be killed, along with those cells that acquired DNA but failed to integrate it, as the episomal plasmid DNA will be lost over the course of several days. Thus, the only cells that remain viable are those that have stably acquired the newly introduced DNA. An exception is when the DNA is in the form of a viral vector that can replicate in the cells.

Numerous vectors are available that contain one of these selectable markers as well as cloning sites for insertion of an additional gene. However, it is also possible to cotransfect the selectable marker *in trans*, as most cells will acquire both plasmids. Plasmid vectors that can be used to establish stable transfectants by either strategy can be purchased from many different commercial vendors; a partial compendium, with links, is found at (**http:// vectordb.atcg.com**). Alternatively, some are available from the noncommercial American Type Culture Collection (ATTC) at (**http://www.atcc.org**).

GENERAL TROUBLESHOOTING CONSIDERATIONS

Although mammalian cells have been transfected with DNAs from a variety of sources, including total genomic DNA (56), the vast majority of experiments involve DNA sequences that have been subcloned and propagated in *Escherichia coli*. Before initiating any of the procedures described below, it should be noted that highly purified DNA tends to produce the best results, both in terms of transfection efficiency and reproducibility. In particular, prokaryotic and bacteriophage DNA were noted to be somewhat toxic to many eukaryotic cells (69). This observation is due at least in part to the carryover of the endotoxin lipopolysaccharide, which also can contaminate plasmid preparations (9,64). Reliable methods of plasmid DNA purification are offered in Chapter 2, although other procedures may prove adequate, especially if high transfection efficiencies are not needed. Additional purification on gels or by other methods is not normally necessary; it is usually easier to replace a poor preparation than it is to clean it up. Some workers subject the DNA sample to precipitation with ethanol and resuspension under sterile conditions. In our experience, this is not necessary, but in the event of bacterial contamination of experiments, such a pretreatment might be helpful.

Finally, although transfection methods have been critical in advancing our understanding of gene expression and function, we must consider the effect of the transfection procedure itself on the state of the cell. As mentioned, most of the procedures result in some loss of viability. However, other effects can be more subtle. Treatment of cells with $CaPO_4$–DNA precipitates, but not DEAE dextran precipitates, resulted in the induction of several genes that are normally responsive to α- or γ-interferon (49). Although the mechanism of action was not fully elucidated, there did not seem to be a requirement for the de novo synthesis of interferons. In contrast, a similar induction of interferon-regulated genes by cationic liposomes apparently was mediated by the activation of the β-interferon gene (32). One can guard against unexpected side effects by performing appropriate controls, such as treating cells with transfection agent in the absence of DNA. If an unwanted effect is observed, another transfection method should be tested.

PROCEDURES

Protocol for DEAE Dextran-Mediated Transfection (39)

1. Plate cells at 2 to 3×10^5 cells/35-mm well (e.g., in a 6-well plate) in normal growth medium one day prior to transfection.
2. Mix 3 to 5 mg of plasmid DNA with 0.2 mL DEAE dextran (0.1–1.0 mg/mL in Hanks buffered saline [HBSS]) in a polystyrene tube.
3. Wash monolayer once with phosphate-buffered saline (PBS). Remove and apply DNA–DEAE dextran mixture; incubate for 30 minutes at 37°C.
4. Add 2.0 mL of growth medium containing serum; incubate for 2 to 3 hours at 37°C.
5. Replace with 2.0 mL fresh growth medium. Incubate for 2 to 3 days prior to harvest.

Reagents

DEAE dextran (500 000 MW; Sigma, St. Louis, MO, USA) is dissolved in HBSS at 1.0 mg/mL then autoclaved. Store stock in 50- to 100-mL aliquots at -20°C. These are thawed as needed to prepare working aliquots and diluted with HBSS if required. Aliquots (1.5 mL) are stored in microfuge tubes at -20°C, and thawed as needed; any excess is discarded.

PBS: Dissolve the following in deionized water, adjust to 1 L, and autoclave.
0.8 g NaCl
0.4 g KCl
60 mg $Na_2HPO_4 \cdot 2H_2O$
60 mg KH_2PO_4
1.0 g Glucose
0.35 g $NaHCO_3$

Notes and Troubleshooting Tips

Steps 1 and 2. Plating cells and preparation of the DNA–DEAE dextran mixture.

Optimization of several parameters should be considered for each cell type. These include the number of cells plated, the amount of DNA transfected, the concentration of DEAE dextran, and the precise times of incubation. Optimization of DEAE dextran concentration is recommended strongly. Use of purified DNA is recommended for optimal results, although the procedure is tolerant of less purified material, such as miniprep DNA.

Step 4. Incubation of cells with DNA mixture.

It is useful to examine cultures at intervals for signs of toxicity. Optimal transfection efficiency is generally associated with some cell death, but excessive death indicates that the chosen conditions are too harsh. Try reducing DEAE concentration or incubation times.

A brief DMSO shock treatment might improve transfection efficiency for many cell types (35,60); see the section below entitled Protocol for $CaPO_4$-Mediated Transfection for conditions. Alternatively, treatment with chloroquine has been reported to enhance expression of transfected genes (37). Chloroquine deacidifies lysosomes, which is thought to reduce degradation of exogenous DNA. Inclusion of chloroquine at 80 to 100 μM in the medium at step 4 may improve expression of transfected DNA.

Comments

DEAE dextran-mediated transfection is quite simple and inexpensive. The procedure may be preferable to $CaPO_4$-mediated transfection for transient expression experiments (35). The tolerance for DNA of lower purity may make the procedure useful for rapid screening of large numbers of constructs.

Protocol for $CaPO_4$-Mediated Transfection (24,56)

1. Plate cells at approximately 3 to 5×10^5 cells/60-mm diameter dish. Incubate overnight in growth medium.

2. Add DNA (total 5–10 μg) to 0.5 mL HEPES-buffered saline (HBS) in bottom of 4.0-mL clear, polystyrene tube.

3. Add 30 μL of 2 M $CaCl_2$ (final concentration of 0.125 M) and mix gently. Let stand at room temperature for 30 to 60 minutes. A very fine precipitate will form, resulting in a slightly cloudy (milky) appearance.

4. Aspirate medium from cells and apply DNA mixture. Let sit at room temperature for 20 minutes, rocking dish once or twice to distribute material.

5. Add 4.5 mL of growth medium with serum and incubate at 37°C for 4 hours.

6. Remove medium with DNA and wash plate once with 2.0 mL fresh growth medium.

7. Apply 1.0 mL of 20% DMSO in HBS. Leave for 4 minutes at room temperature.

8. Aspirate DMSO mixture, wash cells once with 2 to 3 mL medium, re-

place with 5.0 mL growth medium, and incubate overnight.

9. The following day, either replace medium with fresh medium and continue incubation, or split cells for further studies.

Reagents

HBS:
137 mM NaCl
5 mM KCl
0.7 mM Na_2HPO_4
6 mM Dextrose
20 mM HEPES

Adjust pH to between 6.9 and 7.1. Sterilize by filtration through a 0.45-μm membrane.

2 M $CaCl_2$: Sterilize by filtration as above. Make fresh if any precipitate forms.

Notes and Troubleshooting Tips

Step 1. Plating cells.
Optimal cell density may vary. The goal is that the cells are approximately 50% confluent on the day of transfection.
Steps 2 and 3. Formation of the DNA–$CaPO_4$ precipitate.
The pH of the HBS solution is critical. A pH below 6.95 will tend to prevent formation of the DNA precipitate; conversely, a pH above pH 7.1 will promote the formation of large precipitates that are not effective. In addition, $CaCl_2$ stocks should be remade if unused for a period of time (> 6 months).

The total amount of DNA per precipitation should not be reduced below suggested levels. Carrier DNA such as sheared salmon or herring sperm DNA can be used to maintain total DNA concentration. Repurification of plasmid DNA should be considered if sporadic problems with the technique occur.
Step 4. Modifications for cells in suspension.
The above protocol assumes that cells are growing in monolayer culture. However, the procedure has been adapted to cells in suspension (8). After DNA–$CaPO_4$ precipitates are prepared, cells are pelleted by centrifugation (approximately $500\times g$ for 10 min) and then resuspended in the DNA mixture. After incubation at room temperature for 15 to 30 minutes to allow adsorption of DNA to the cells, the cells are diluted into growth medium and transferred to culture dishes or flasks for further incubation.
Step 5. Incubation of cells with DNA.

Cells should be monitored periodically, as significant toxicity may become apparent at this stage. Decreasing exposure time may limit cell damage. Some investigators incubate under conditions of reduced CO_2 so as to minimize pH changes.

Step 7. Posttreatment with DMSO.

The DMSO shock (step 7) can be omitted, although some reduction in efficiency may be observed. If toxicity is the problem, less DMSO can be used, or exposure time may be decreased. Alternatively, substitution with 15% glycerol in HBS can be tried. Optimal conditions are highly cell-type dependent.

Comments

CaPO$_4$-mediated transfection has been used for many applications, especially for the isolation of stably transfected cells. Although the cost is low, preparation of the reagents can be finicky, and not all cells are efficiently transfected by this method. Premade solutions are commercially available from a number of manufacturers (the list of vendors in Table 1 can serve as a starting point) and are worth investigating if limited use of the procedure is anticipated.

Protocol for Lipid-Mediated DNA Uptake

1. Plate cells at 2 to 3×10^5 cells/35-mm well. Incubate overnight in growth medium.
2. Dilute 1 to 5 µg DNA into a total of 100 µL serum-free medium in a polystyrene tube.
3. Dilute 5 to 20 µL lipid separately into 100 µL serum-free medium in a polystyrene tube.
4. Combine the two solutions and leave at room temperature for 30 minutes.
5. Wash cells with serum-free medium, then add DNA mixture after adjusting volume to 1.0 mL total with serum-free medium.
6. Incubate for 4 to 24 hours at 37°C.
7. Remove DNA mixture and replace with fresh growth medium for further incubation.

Notes and Troubleshooting Tips

Step 1. Plating cells.

Plating density will need to be determined for each cell type. Generally, cells should be approximately 70% to 80% confluent on the day of transfection. The procedure may be scaled up or down as convenient, but the 35-mm format (6-well plate) represents a convenient compromise between obtaining sufficient material to analyze and reducing the amount of reagents and DNA used.

<u>Steps 2 through 4.</u> Preparation of the DNA–lipid complexes.

The optimal ratio of DNA:lipid will need to be determined in preliminary experiments. The lipid should be added to the medium directly to avoid losses to the surface of the tube. The choice of polystyrene also minimizes loss to the surface.

<u>Steps 5 and 6.</u> Incubation of cells with DNA.

The optimal incubation time will vary with cell type and must be determined empirically. At least one serum-free medium, OPTI-MEM™ (Life Technologies, Rockville, MD, USA) has been formulated for use in these transfection procedures, and a number of vendors suggest that use of this product may improve transfection efficiency. Some lipid reagents are reported to be tolerant of at least 5% fetal bovine serum. Should the absence of serum for a period of a few hours prove traumatic for the cells of interest, this will be a factor in the choice of lipid reagent.

It is generally advisable to expose the cells to the lipolexes in as small a volume as is practical in order to maximize exposure to the cell surface. Alternatively, centrifugation of lipoplexes onto adherent cells can improve transfection efficiency (63).

Comments

This procedure has been adapted from that provided with the LIPOFECT-AMINE™ reagent (Life Technologies), but most other protocols for other agents are quite similar and resemble those originally described for the cationic lipid DOTMA (13).

Lipid-mediated transfections have been gaining in popularity for many applications, and have proven very useful for working with a number of cell types that were refractory to other methods. Vendors of lipid reagents tend to track the use of their products, and thus optimal lipid composition for use with a given cell type may be known, at least within that manufacturer's product line. If no information is available, several companies with more than one product offer sampler packs to facilitate optimization of transfection conditions. These packages may be attractive if conditions will need to be optimized for more than one cell line.

Another application that has received considerable attention is to use lipoplexes for in vivo gene delivery (55,62). However, evidence suggests that optimal conditions for gene delivery to cells in culture are not identical to those optimal for in vivo delivery, due to clearance of the DNA-containing complexes. In addition, effectiveness will depend on other parameters such as stability and the tendency of certain lipids to target to specific organs. This is a very active area of research, and new compounds and formulations appear with high frequency.

Protocol for PEI-Mediated DNA Uptake (3)

1. Plate cells at 5×10^4 cells/10-mm well (e.g., 24-well plate) one day prior to transfection.

2. Immediately before transfection, rinse cells with serum-free medium, then add 1 mL serum-free medium/well.

3. Dilute 2 μg plasmid DNA into 50 μL with 150 mM NaCl.

4. Dilute desired amount PEI stock into 50 μL with 150 mM NaCl. Incubate at 25°C for 10 minutes.

5. Add diluted PEI solution dropwise to the DNA solution, vortex mix, and incubate for 10 minutes at 25°C.

6. Add mixture to cells and incubate are 37°C. After 3 to 4 hours, add serum as needed and continue to incubate cells as desired.

Reagents

PEI stock:
9 mg of a 50% (wt/vol) solution of PEI (M_r = 25 or 800 kDa; from Sigma or Fluka Chemical [Milwaukee, WI, USA]) is dissolved in 10 mL water and neutralize with HCl (3). Filter-sterilize prior to use.

Notes

The amount of PEI to be added may need to be determined empirically. However, use of 5 to 6 μL of the PEI stock should yield a ratio of approximately 9 PEI nitrogens per DNA phosphate, the optimal ratio for mouse 3T3 fibroblasts (3). These same workers also found that the addition of the polymer solution to the plasmid was much more effective than reversing the order of addition. Other workers have reported successful transfection using PEI of average M_r = 25 kDa. Transfection quality PEI is also supplied as ExGen™ 500 (MBI Fermentas, Amherst, NY, USA).

Comments

PEI and other polymers such as dendrimers (commercially available as SuperFect™ from Qiagen [Valencia, CA, USA]) represent additional options for cells that prove hard to transfect. Many of these materials are being investigated for their ability to deliver genes in intact animals and humans, an application that presents additional challenges, as described above.

Protocol for Electroporation

1. Harvest the cells. If cells are in suspension, transfer them to a conical centrifuge tube and pellet at approximately $500\times g$. If cells are adherent, trypsinize to release from the surface, then add serum, and proceed as for suspension cells.
2. Wash once with PBS and resuspend in PBS at a concentration of approximately 10^7 cells/mL. Transfer 0.5 mL cells to a sterile electroporation cuvette on ice.
3. Add DNA (in Tris-EDTA or water) to the cells at a final concentration of 10 to 100 µg/mL. Mix gently and leave on ice for 10 minutes.
4. Charge the electroporation device as indicated by the manufacturer and apply desired pulse.
5. Let cells and DNA incubate on ice for 10 minutes.
6. Transfer cells to a 10-cm tissue culture dish or T75 flask in growth medium plus serum and incubate as desired.

Notes and Troubleshooting Tips

<u>Step 1.</u> Collection of cells.
Cells should be actively growing when collected for electroporation. It is convenient to subculture cells 1 to 2 days prior to an intended experiment to a density that should result in cell proliferation. HEPES-buffered saline may be substituted for PBS if desired; if cells are especially fragile, growth medium without serum can be used. However, expect to adjust voltage parameters if ionic conditions are changed.

<u>Step 2.</u> Transfer of cells to cuvette.
Although manufacturers tend to recommend only a single use of disposable cuvettes for electroporation, they can be reused after careful rinsing in water and soaking in ethanol to sterilize the chamber. Dry completely prior to use. Carefully inspect the electrode surfaces for signs of wear or evidence of arcing and discard any that appear imperfect. If there are concerns regarding cross-contamination of DNAs, however, use of new cuvettes is advised.

<u>Step 4.</u> Typical voltages for electroporation under the high resistance conditions described are 600 to 800 kV/cm, with a capacitor setting of 500 to 1000 µF. A trial experiment varying the two parameters separately should be performed for each cell type to determine the optimal conditions.

<u>Step 5.</u> DNA uptake.

Incubation both before and after electroporation can be performed at room temperature if the cells do not tolerate the cold well.

Comments

Electroporation is a useful alternative for cells that are refractory to chemical means of transfection. The protocol offered above should be considered merely a starting point, as the variables are several, and machines from different sources may have different settings available. However, at least two vendors of electroporation apparatus have collected many protocols that are available at the Web sites (**www.biorad.com**) and (**www.btxonline.com**). These protocols are collected from researchers and offer specific conditions that have proven successful with individual cell types. Conditions may need to be modified to adapt a procedure to a different piece of equipment, but the compendiums represent a useful resource.

If the desired result is the selection of stably transfected cells, DNA should be linearized prior to mixing with the cells (51). For transient transfection, supercoiled DNA is likely to remain more stable and thus result in higher levels of gene expression.

Measuring Transfection Efficiency by X-Gal Staining (54)

1. Wash adherent cell monolayer once with 1.0 mL PBS.
2. Add 1.0 mL of fixative and leave at 4°C for 5 minutes.
3. Remove fixative and wash 2 to 3 times with 1.0 mL of PBS.
4. Add 1.0 mL stain and incubate at 37°C for 1 to 24 hours.

Reagents

PBS: See the section entitled Protocol for DEAE Dextran-Mediated Transfection.

Fixative: Dilute 37% formaldehyde solution 1:10 in PBS.

Stain: Prepare from stock solutions.
8.7 mL PBS
1.0 mL 10× Fe salts (10× stock: 50 mM potassium ferricyanide, 50 mM potassium ferrocyanide)
20 µL 1.0 M $MgCl_2$
250 µL X-gal (Stock: 40 mg/mL in dimethyl formamide; store at -20°C)

Notes and Troubleshooting Tips

<u>Steps 1 through 3.</u> Fixation.

Volumes given are suitable for 35- to 60-mm wells and should be adjusted for smaller or larger dishes. Different protocols call for the substitution of glutaraldehyde for formaldehyde, or a combination of both; convenience is the primary consideration in most instances. Although the procedure can be adapted for suspension cells, use of a fluorescent substrate and FACS analysis is likely to prove simpler (16,46).

<u>Step 4.</u> Staining.

Volumes given are suitable for 35- to 60-mm wells and should be adjusted for smaller or larger dishes. Stain solution will keep at 4°C for up to one week, and stocks are stable longer. Progress of staining should be monitored at intervals using either brightfield or phase-contrast microscopy. Ideally, cells will be incubated until no additional blue cells appear. However, some cell types begin to stain very lightly with prolonged incubation because of endogenous enzyme activities. This can be identified by staining control untransfected cells in parallel; maintaining the pH of the stain solution above 7.0 may also help improve signal to noise.

Protocol for Stable Transfections Using G418 Selection

1. Include the selectable marker gene in the transfection. This can be supplied in *cis* on the same plasmid as the gene of interest, as in a number of commercially available vectors. Alternatively, the marker gene can be co-transfected with the plasmid of interest. Usually the latter is included at a 3- to 10-fold molar excess over the selectable plasmid.
2. Perform transfection using one of the methods described above. In some cases, linearization of plasmid DNA may enhance integration efficiency. However, this may not be feasible, depending on features of the individual construct.
3. One to three days posttransfection, cells should be transferred to selective medium. Add G418 (geneticin) to the growth medium to achieve a final concentration of 200 to 1000 µg/mL active drug. For each cell type, perform a dose response curve and monitor cell viability. Chose the lowest dose at which all cells are killed, but note that complete killing may take 2 weeks, depending on cell type. If cultures are very dense, replating at a lower cell density may be desirable, especially if individual colonies will be picked.
4. Once all nontransfected cells have died (parallel control cultures are useful in determining this point), two options are available. Either the entire population may be expanded or single clones may be picked. In the

former case, trypsinize the cells as usual and replate, maintaining drug selection. In the latter case, use a microscope to mark the boundaries of individual clonal outgrowths, with the goal of choosing healthy well-isolated colonies. Cloning cylinders may be used to trypsinize colonies individually; released cells are then transferred into wells containing fresh growth medium plus drug. Alternatively, a convenient solution is to saturate small rounds of filter paper (sterilized by autoclaving) with trypsin. Lay one filter over each colony, wait several minutes, then transfer the filter, cell side down, to a single 10-mm well containing fresh growth medium plus drug (43).

Reagents

G418: Dissolve drug in PBS or HBS at 100 mg/mL (active drug concentration; refer to activity specifications for each lot individually, but activity is typically 0.5 mg active drug/mg powder). Filter-sterilize. Stock solution may be stored frozen.

ACKNOWLEDGMENTS

The author acknowledges support from the Division of Gastroenterology and Hepatology at Thomas Jefferson University and from the National Institute of Arthritis and Musculoskeletal and Skin Diseases (AR42887).

REFERENCES

1. **Andreason, G.L. and G.A. Evans.** 1988. Introduction and expression of DNA molecules in eukaryotic cells by electroporation. BioTechniques 6:650-660.
2. **Bandyopadhyay, P., B.T. Kren, X. Ma, and C.J. Steer.** 1998. Enhanced gene transfer into HuH-7 cells and primary rat hepatocytes using targeted liposomes and polyethylenimine. BioTechniques 25:282-292.
3. **Boussif, O., F. Lezoualc'h, M.A. Zanta, M.D. Mercny, D. Scherman, B. Demeneix, and J.-P. Behr.** 1995. A versatile vector for gene and oligonucleotide transfer into cells in culture and in vivo: polyethylenimine. Proc. Natl. Acad. Sci. USA 92:7297-7301.
4. **Cappechi, M.R.** 1980. High efficiency transformation by direct microinjection of DNA into cultured mammalian cells. Cell 22:479-488.
5. **Chalfie, M., Y. Tu, G. Euskirchen, W.W. Ward, and D.C. Prasher.** 1994. Green fluorescent protein as a marker for gene expression. Science 263:802-805.
6. **Chen, C. and H. Okayama.** 1988. Calcium phosphate-mediated gene transfer: a highly efficient system for stably transforming cells with plasmid DNA. BioTechniques 6:632-638.
7. **Chu, G., H. Hayakawa, and P. Berg.** 1987. Electroporation for the efficient transfection of mammalian cells with DNA. Nucleic Acids Res. 15:1311-1326.
8. **Chu, G. and P.A. Sharp.** 1981. SV40 DNA transfection of cells in suspension: analysis of the efficiency of transcription and translation of T-antigen. Gene 13:197-202.
9. **Cotten, M., A. Baker, M. Saltik, E. Wagner, and M. Buschle.** 1994. Lipopolysaccharide is a frequent

contaminant of plasmid DNA preparations and can be toxic to primary human cells in the pesence of adenovirus. Gene Ther. *1*:239-246.

10. **Cotten, M., E. Wagner, and M.L. Birnstiel.** 1993. Receptor-mediated transport of DNA into eukaryotic cells. Methods Enzymol. *217*:618-644.

11. **Farhood, H., N. Serbina, and L. Huang.** 1995. The role of phosphatidylethanolamine in cationic liposome meditiated gene transfer. Biochim. Biophys. Acta *1235*:289-295.

12. **Felgner, P.L., Y. Barenholz, J.P. Behr, S.H. Cheng, P. Cullis, L. Huang, J.A. Jessee, L. Seymour, F. Szoka, A.R. Thierry, et al.** 1997. Nomenclature for synthetic gene delivery systems. Hum. Gene Ther. *8*:511-512.

13. **Felgner, P.L., T.R. Gadek, M. Holm, R. Roman, H.W. Chan, M. Wnez, J.P. Northrop, G.M. Ringold, and M. Danielsen.** 1987. Lipofection: a highly efficient, lipid-mediated DNA-transfection procedure. Proc. Natl. Acad. Sci. USA *84*:7413-7417.

14. **Felgner, J.H., R. Kumar, C.N. Sridhar, C.J. Wheeler, Y.J. Tsai, R. Border, P. Ramsey, M. Martin, and P.L. Felgner.** 1994. Enhanced gene delivery and mechanism studies with a novel series of cationic lipid formulations. J. Biol. Chem. *269*:2550-2561.

15. **Felgner, P.L. and G.M. Ringold.** 1989. Cationic liposome-mediated transfection. Nature *337*:387-388.

16. **Fiering, S.N., M. Roederer, G.P. Nolan, D.R. Micklem, D.R. Parks, and L.A. Herzenberg.** 1991. Improved FACS-Gal: flow cytometric analysis and sorting of viable eukaryotic cells expressing reporter gene constructs. Cytometry *12*:291-301.

17. **Fraley, R., R.M. Straubinger, G. Rule, E.L. Springer, and D. Papahadjopoulos.** 1981. Liposome-mediated delivery of deoxyribonucleic acid to cells: enhanced efficiency of delivery related to lipid composition and incubation conditions. Biochemistry *20*:6978-6987.

18. **Freshney, R.I.** 1986. Introduction: principles of sterile technique and cell propagation, p. 1-11. *In* R.I. Freshney (Ed.), Animal Cell Culture: A Practical Approach. IRL Press, Oxford.

19. **Friend, D.S., D. Papahadjopoulos, and R.J. Debs.** 1996. Endocytosis and intracellular processing accompanying transfection mediated by cationic liposomes. Biochim. Biophys. Acta *1278*:41-50.

20. **Gaiano, N. and N. Hopkins.** 1996. Introducing genes into zebrafish. Biochim. Biophys. Acta *1288*:O11-O14.

21. **Gao, X. and L. Huang.** 1996. Potentiation of cationic liposome-mediated gene delivery by polycations. Biochemistry *35*:1027-1036.

22. **Gershon, H., R. Ghirlando, S.B. Guttman, and A. Minsky.** 1993. Mode of formation and structural features of DNA-cationic liposome complexes used for transfection. Biochemistry *32*:7143-7151.

23. **Godbey, W.T., K.K. Wu, and A.G. Mikos.** 1999. Tracking the intracellular path of poly(ethylenimine)/DNA complexes for gene delivery. Proc. Natl. Acad. Sci. USA *96*:5177-5181.

24. **Graham, F.L. and A.J. van der Eb.** 1973. A new technique for the assay of infectivity of human adenovirus 5 DNA. Virology *52*:456-467.

25. **Griffiths, T., M. Russell, K. Froning, B.D. Brown, S.M. Scanlon, M. Almazan, R. Marcil, and J.P. Hoeffler.** 1997. The PerFect™ lipid optimizer kit for maximizing lipid-mediated transfection of eukaryotic cells. BioTechniques *22*:982-987.

26. **Haensler, J. and F.C. Szoka, Jr.** 1993. Polyamidoamine cascade polymers mediate efficient transfection of cells in culture. Bioconjug. Chem. *4*:372-372.

27. **Hirota, S., C. Tros de Ilarduya, L.G. Barron, and F.C. Szoka, Jr.** 1999. Simple mixing device to reproducibly prepare cationic lipid-DNA complexes (lipoplexes). BioTechniques *27*:286-290.

28. **Koltover, I., T. Salditt, J.O. Rädler, and C.R. Safinya.** 1998. An inverted hexagonal phase of cationic liposome-DNA complexes related to DNA release and delivery. Science *281*:78-81.

29. **Kukowska-Latallo, J.F., A.U. Bielinska, J. Johnson, R. Spindler, D.A. Tomalia, and J.R. Baker, Jr.** 1996. Efficient transfer of genetic material into mammalian cells using Starburst polyamidoamine dendrimers. Proc. Natl. Acad. Sci. USA *93*:4897-4902.

30. **Lechardeur, D., K.J. Sohn, M. Haardt, P.B. Joshi, M. Monck, R.W. Graham, B. Beatty, J. Squire, H. O'Brodovich, and G.L. Lukacs.** 1999. Metabolic instability of plasmid DNA in the cytosol: a potential barrier to gene transfer. Gene Ther. *4*:482-497.

31. **Legendre, J.Y. and F.C. Szoka, Jr.** 1992. Delivery of plsamid DNA into mammalian cell lines using pH-sensitive liposomes: comparison with cationic liposomes. Pharm. Res. *9*:1235-1242.

32. **Li, X.-L., M. Boyanapalli, X. Weihua, D.V. Kalvakolamy, and B.A. Hassel.** 1998. Induction of interferon synthesis and activation of interferon-stimulated genes by liposomal transfection reagents. J.

Interferon Cytokine Res. *18*:947-952.

33.Litzinger, D.C. and L. Huang. 1992. Phosphatidylethanolamine liposomes: drug delivery, gene transfer and immunodiagnostic applications. Biochim. Biophys. Acta *1113*:210-227.

34.Loeffler, J.-P. and J.-P. Behr. 1993. Gene transfer into primary and established mammalian cell lines with lipopolyamine-coated DNA. Methods Enzymol. *217*:599-618.

35.Lopata, M.A., D.W. Cleveland, and B. Sollner-Webb. 1984. High level transient expression of a chloramphenicol acetyl transferase gene by DEAE-dextran mediated DNA transfection coupled with a dimethyl sulfoxide or glycerol shock treatment. Nucleic Acids Res. *12*:5707-5717.

36.Loyter, A., G.A. Scangos, and F.H. Ruddle. 1982. Mechanisms of DNA uptake by mammalian cells: fate of exogenously added DNA monitored by the use of fluorescent dyes. Proc. Natl. Acad. Sci. USA *79*:422-426.

37.Luthman, H. and G. Magnusson. 1983. High efficiency polyoma DNA transfection of chloroquine treated cells. Nucleic Acids Res. *11*:1296-1308.

38.Matsui, H., L.G. Johnson, S.H. Randell, and R.C. Boucher. 1997. Loss of binding and entry of liposome-DNA complexes decreases transfection efficiency in differentiated airway epithelial cells. J. Biol. Chem. *272*:1117-1126.

39.McCutchan, J.H. and J.S. Pagano. 1968. Enhancement of the infectivity of simian virus 40 deoxyribonucleic acid with dietylaminoethyl-dextran. J. Natl. Cancer Inst. *41*:351-356.

40.Miller, A.D., D.G. Miller, J.V. Garcia, and C.M. Lynch. 1993. Use of retroviral vectors for gene transfer and expression. Methods Enzymol. *21*:581-599.

41.Morgan, B.A. and D.M. Fekete. 1996. Manipulating gene expression with replication-competent retroviruses. Methods Cell Biol. *51*:185-220.

42.Mortimer, I., P. Tam, I. MacLachlan, R.W. Graham, E.G. Saravolac, and P.B. Joshi. 1999. Cationic lipid-mediated transfection of cells in culture requires mitotic activity. Gene Ther. *4*:403-411.

43.Moser, M.J. 1998. Re: Cloning cylinders: try paper discs instead. **http://www.bio.net/hypermail/ fluorpro/fluorpro.199806/0012.html**

44.Mounkes, L.C., W. Zhong, G. Cipres-Palacin, T.D. Heath, and R.J. Debs. 1998. Proteoglycans mediate cationic liposome-DNA complex-based gene delivery in vitro and in vivo. J. Biol. Chem. *273*:26164-26170.

45.Neumann, E., M. Schaefer-Ridder, Y. Wang, and P.H. Hofschneider. 1982. Gene transfer into mouse lyoma cells by electroporation in high electric fields. EMBO J. *1*:841-845.

46.Nolan, G.P., S. Fiering, J.-F. Nicolas, and L.A. Herzenberg. 1988. Fluorescence-activated cell analysis and sorting of viable mammalian cells based on β-D-galactosidase activity after transduction of *Escherichia coli lacZ*. Proc. Natl. Acad. Sci. USA *85*:2603-2607.

47.Palmer, T.D., R.A. Hock, W.R.A. Osborne, and A.D. Miller. 1987. Efficient retrovirus-mediated transfer and expression of a human adenosine deaminase gene in diploid skin fibroblasts from an adenosine-deficient human. Proc. Natl. Acad. Sci. USA *84*:1055-1059.

48.Pellicer, A., D. Robins, B. Wold, R. Sweet, J. Jackson, I. Lowy, J.M. Roberts, G.K. Sim, S. Silverstein, and R. Axel. 1980. Altering genotype and phenotype by DNA-mediated gene transfer. Science *209*:1414-1422.

49.Pine, R., D.E. Levy, N. Reich, and J.E. Darnell, Jr. 1988. Transcriptional stimulation by CaPO$_4$-DNA precipitates. Nucleic Acids Res. *16*:1371-1378.

50.Pollard, H., J.-S. Remy, G. Loussouarn, S. Demolombe, J.-P. Behr, and D. Escande. 1998. Polyethylenimine but not cationic lipids promotes transgene delivery to the nucleus in mammalian cells. J. Biol. Chem. *273*:7507-7511.

51.Potter, H. 1988. Electroporation in biology: methods, applications and instrumentation. Anal. Biochem. *174*:361-373.

52.Potter, H., L. Weir, and P. Leder. 1984. Enhancer-dependent expression of human k immunogobulin genes introduced into mouse pre-B lymphocytes by electroporation. Proc. Natl. Acad. Sci. USA *91*:7161-7165.

53.Rädler, J.O., I. Koltover, T. Salditt, and C.R. Safinya. 1997. Structure of DNA-cationic liposome complexes: DNA intercalation in multilamellar membranes in distinct interhelical packing regimes. Science *275*:810-814.

54.Sanes, J.R., J.L.R. Rubenstein, and J.-F. Nicolas. 1986. Use of a recombinant retrovirus to study post-implantation cell lineage in mouse embryos. EMBO J. *5*:3133-3142.

55.Scherman, D., M. Bessodes, B. Cameron, J. Herscovici, H. Hofland, B. Pitard, F. Soubrier, P. Wils, and J. Crouzet. 1998. Application of lipids and plasmid design for gene delivery to mammalian cells. Curr. Opin. Biotechnol. *9*:480-485.

56. Shih, C., B.-Z. Shilo, M.P. Goldfarb, A. Dannenberg, and R.A. Weinberg. 1979. Passage of phenotypes of chemically transformed cells via transfection of DNA and chromatin. Proc. Natl. Acad. Sci. USA *76*:5714-5718.

57. Smith, J.G., R.L. Walzem, and J.B. German. 1993. Liposomes as agents of DNA transfer. Biochim. Biophys. Acta *1154*:327-340.

58. Southern, P.J. and P. Berg. 1982. Transformation of mammalian cells to antibiotic resistance with a bacterial gene under control of the SV40 early region promoter. J. Mol. Appl. Genet. *1*:327-341.

59. Stow, N.D. and N.M. Wilkie. 1976. An improved technique for obtaining enhanced infectivity with herpes simplex virus type 1 DNA. J. Gen. Virol. *33*:447-458.

60. Sussman, D.J. and G. Milman. 1984. Short-term, high-efficiency expression of transfected DNA. Mol. Cell. Biol. *4*:1641-1643.

61. Tseng, W.C., F.R. Haselton, and T.D. Giorgio. 1997. Transfection by cationic liposomes using simultaneous single cell measurements of plasmid delivery and transgene expression. J. Biol. Chem. *272*:25641-25647.

62. Tseng, W.C. and L. Huang. 1998. Liposome-based gene therapy. Pharm. Sci. Tech. Today *1*:206-213.

63. Verma, R.S., D. Giannola, W. Shlomchik, and S.G. Emerson. 1998. Increased efficiency of liposome-mediated transfection by volume reduction and centrifugation. BioTechniques 25:46-49.

64. Weber, M., K. Möller, M. Welzeck, and J. Schorr. 1995. Effects if lipopolysaccharide on transfection efficiency in eukaryotic cells. BioTechniques *19*:930-940.

65. Wigler, M., R. Sweet, G.K. Sim, B. Wold, A. Pellicer, E. Lacy, T. Maniatis, S. Silverstein, and R. Axel. 1979. Transformation of mammalian cells with genes from procaryotes and eucaryotes. Cell *16*:777-785.

66. Wilson, T., D. Papahadjopoulos, and R. Taber. 1977. Biological properties of poliovirus encapsulated in lipid vesicles: antibody resistance and infectivity in virus-resistant cells. Proc. Natl. Acad. Sci. USA *74*:3471-3475.

67. Wu, G.Y. and C.H. Wu. 1987. Receptor-mediated gene transformation by a soluble DNA carrier system. J. Biol. Chem. *262*:4429-4432.

68. Xu, Y. and F.C. Szoka. 1996. Mechanism of DNA release from cationic liposome/DNA complexes used in cell transfection. Biochemistry *35*:5616-5623.

69. Yoder, J.I. and T. Ganesan. 1983. Procaryotic genomic DNA inhibits mammalian cell transformation. Mol. Cell. Biol. *3*:956-959.

70. Zabner, J., A.J. Fasbender, T. Moninger, K.A. Poellinger, and M.J. Welsh. 1995. Cellular and molecular barriers to gene transfer by a cationic lipid. J. Biol. Chem. *270*:18997-19007.

71. Zanta, M.A., P. Belguise-Valladier, and J.P. Behr. 1999. Gene delivery: a single nuclear localization signal peptide is sufficient to carry DNA to the cell nucleus. Proc. Natl. Acad. Sci. USA *96*:91-96.

72. Zhou, X. and L. Huang. 1994. DNA transfection mediated by cationic liposomes containing lipopolylysine: characterization and mechanism of action. Biochim. Biophys. Acta *1189*:195-203.

6

Generation of Transgenic *Drosophila melanogaster* by *P* Element-Mediated Germline Transformation

Richard M. Cripps[1] and Sanford I. Bernstein[2]
[1]*Department of Biology, University of New Mexico, Albuquerque, NM;* [2]*Department of Biology and Molecular Biology Institute, San Diego State University, San Diego, CA, USA*

INTRODUCTION

Perfection of a method to generate transgenic *Drosophila melanogaster* revolutionized the fields of genetics and developmental biology. The methodology, which involves mobilizing engineered transposable elements to insert genes into the *Drosophila* germline, is remarkably unchanged since its description by Spradling and Rubin in 1982 (44). In this chapter, we outline the theoretical background of transposable element-mediated germline transformation and then discuss the experimental protocols and potential points of variation and difficulty.

Historical Overview of *P* Element-Mediated Transformation

P elements are naturally occurring transposable elements found in populations of *Drosophila melanogaster*, the common fruit fly. Rubin and Spradling harnessed the natural ability of these elements to transpose, and they developed a method for using them to stably transform genes into the *Drosophila* germline (41,44). Given the utility of *Drosophila melanogaster* for genetic and developmental studies, the ability to produce transgenic organisms is an incredible boon to scientists studying gene regulation and function in eukaryotic organisms.

Gene Transfer Methods
Edited by Pamela A. Norton and Laura F. Steel
©2000 Eaton Publishing, Natick, MA

P elements first were described genetically as factors that induce a syndrome called hybrid dysgenesis; the affected organisms display dysgenic traits including high rates of mutation, sterility, male recombination, and chromosome breakage in their germline cells (21,22,36,45). An interesting aspect of this phenomenon is that the affected organism must result from the cross of an *M* strain mother and a *P* strain father. While the resulting hybrid displays the dysgenic traits in its germline, an individual resulting from crossing an *M* strain father and a *P* strain mother does not. Thus the activation of the genetic factors (*P* elements) is dependent upon the presence of the factors in the hybrid as well as the so-called *M* cytotype contributed by the mother.

Molecular cloning techniques permitted the isolation and characterization of the DNA elements that cause hybrid dysgenesis (41). The full-size *P* element is nearly 3 kb in length and contains 31-bp inverted repeats at its ends (Figure 1); insertion of the element results in production of an 8-bp direct repeat at the target site (34). In addition to full-size elements, shorter elements are capable of transposition as well. They retain the inverted terminal repeats, but have internal deletions.

The *P* element encodes a transposase that is capable of recognizing the terminal repeats of other *P* elements and transposing them to new locations. Transposition occurs by a cut-and-paste mechanism (4,5), wherein the donor DNA is cut at specific sites near the terminal repeats and inserted into the target site. The transposase produces a single-stranded terminal region at the *P* element ends. Transfer of the element and repair of the target site require

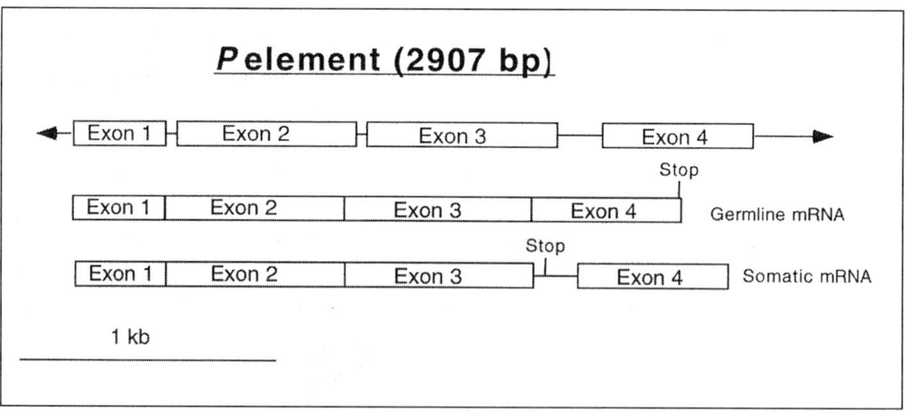

Figure 1. Structure and transcription of the *P* element. The element is 2.9 kb and terminates in 31-bp inverted repeats (arrowheads). Direct repeats of 8 bp arise in the genomic DNA as a result of *P* element insertion. The four exons encode transposase, and this 87-kDa protein is produced in the germline as a result of RNA splicing to remove three intervening sequences (Germline mRNA). In somatic cells, the third intron is retained (Somatic mRNA), resulting in premature translation termination. This yields a 66-kDa protein that is not functional as a transposase. Translation termination signals are indicated as "Stop."

endogenous DNA repair enzymes.

The transposase is encoded by four open reading frames within the full-size *P* element (Figure 1). The germline specificity of transposase function is dictated by alternative RNA splicing (24). The removal of the 3′-most intron only occurs in germline tissue to yield a functional transposase. The mechanism of germline-specific removal of this intron has been studied in some detail and arises from a splicing inhibitor present in somatic tissue and absent from germline cells (1). Production of transposase in somatic tissues by pre-splicing of the two terminal exons results in high mobility of integrated *P* elements (24). Such constructs that express functional transposase often are used to mobilize *P* elements in vivo (37,38).

Repression of transposase function in the germline of *P* strains is a result of both repression of transcription of the element and reduced levels of removal of the terminal intron (39). The 66-kDa protein produced as a result of premature translational termination acts as a repressor of *P* element transposition, as do some proteins derived from *P* elements with internal deletions (32,37). Repression of transposition results when the truncated proteins compete with functional transposase at the terminal binding sites of the *P* element targets (25).

The methodology described in this chapter for using *P* elements to transform the *Drosophila* germline is similar to that initially developed by Rubin and Spradling in the early 1980s (44). The DNA to be transposed is first cloned between the inverted repeat sequences of the *P* element, so that the transposase will recognize and transpose it. A marker gene is usually included in the construct to allow ready identification of transformants in the offspring of the injected individuals. Young *Drosophila* embryos are collected from population cages, their outer chorion membranes are removed, the embryos are briefly desiccated, and a solution containing the DNA is injected into the posterior of the embryo (near the developing germ cells) using a micropipet. The embryos are injected prior to cellular blastoderm formation, before cell membrane formation that might inhibit DNA uptake. DNA encoding functional transposase must be co-injected, so that the transposase protein will be produced in the germline cells. This allows recognition of the *P* element ends surrounding the cloned gene and insertion of this cassette into the chromosomal DNA. To prevent insertion of the intact transposase gene in the genome, a modified version of the intact *P* element is used (19). It cannot insert into the genome itself, because it lacks a functional terminal repeat. Continued production of transposase by an intact element inserted in the genome would lead to *P* element mobility rather than stable germline transmission. Once the injected embryos become adults, they are mated to an appropriate fly strain so that the transformed offspring can be identified by the marker contained on the transposed *P* element. Subsequently, lines containing single transgenes are established, and the location of each transgene is determined by standard genetic crosses.

Other Methods for Studying Gene Regulation in *Drosophila* via Embryo Injection

The general methodology used for embryo injection in *P* element-mediated germline transformation has a number of other applications for studying gene regulation. The advantage to these procedures is that data can be collected from the injected individuals, rather than having to establish stable lines. DNA can be injected, and gene expression studied by transient analysis of product accumulation in embryos, larvae, and adults (30). Although results pertinent to transcriptional or posttranscriptional regulatory signals or protein function can be obtained rapidly, genes are not partitioned equally to all cells, and some cells may receive no copies of the injected gene. Thus, multiple individuals need to be analyzed to ensure that conclusions are not based on gene copy number considerations.

Injecting RNA directly into the blastoderm allows investigators to study effects of genes acting early in development. RNA isolated from organisms or in vitro transcribed RNA have been used successfully. This approach can be employed to rescue genetic defects with a wild-type transcript, to demonstrate the function of an engineered protein, or to study the functional roles of particular sequences on the transcript (2,10,27).

Determining the function of an endogenous protein can be approached by injection of antisense RNA. Although full inhibition of endogenous RNA function is dependent upon choice of antisense transcript, partial inhibition can provide insight into gene function during embryo development (40,42). There may be multiple mechanisms by which antisense RNA inhibits protein synthesis, e.g., transcriptional inhibition, translational inhibition, degradation of RNA duplexes.

Recently, procedures for inhibiting gene function by injection of double-stranded RNA (inhibitory RNA or iRNA) have been developed for use in *Drosophila*. iRNA's applicability for disrupting gene expression in the nematode *Caenorhabditis elegans* is well documented, although its mechanism of action is unclear (43,46). The injection into *Drosophila* embryos of double-stranded RNA encoding a number of embryonically expressed proteins generally results in phenotypes that mimic or are identical to genetic null mutants (16,20,31). Lesser effects upon gene expression in adults have been documented as well (31).

Applications of *P* Element-Mediated Transformation

A major application of *P* element-mediated transformation is for the verification that a particular mutant phenotype results from a mutation in a specific gene. A gene encoding the wild-type version of the candidate protein is transformed into the germline using the *P* element technique and demonstrated to rescue the mutant phenotype. In cases in which the transcriptional promoter of the gene in question is used, this constitutes proof that the correct gene has been identified. In cases where an inducible promoter or tissue-specific promoter is used in place of the endogenous promoter, a reasonable case can be made for identification of the mutant gene, given the caveat that expression of a different gene in the programmed temporal or spatial pattern might rescue the mutant defect.

The *P* element transformation procedure is useful for the functional analysis of mutant and chimeric proteins. This is particularly the case when an engineered protein can be expressed in a genetically null background. This approach allows one to determine the role of specific protein domains or subdomains in vivo. Further, the role of transcriptional regulatory elements can be dissected by step-wise mutagenesis of promoter–enhancer sequences and production of transgenic lines. For such experiments, a reporter gene encoding proteins like *Escherichia coli* β galactosidase or jellyfish green fluorescent protein (GFP) is oftentimes fused to the regulatory elements so that gene expression can be readily assayed by tissue staining with a chromogenic substrate (5-bromo-4-chloro-3-indolyl-β-D-galactopyranoside [X-gal] in the case of β-galactosidase) or microscopic observation under UV light (GFP). GFP is particularly advantageous in that its expression can be observed in living organisms. See Timmons et al. (47) for a discussion of the advantages of each of these reporter systems in *Drosophila*.

Inducible promoters (such as a heat shock gene promoter) or tissue-specific promoters are useful for expressing genes introduced by *P* element-mediated transformation. This approach allows genes to be expressed at specific times of development or in particular tissues. Over-expression or ectopic expression of a protein can sometimes yield insight into the role of the protein during its normal expression pattern, as was demonstrated for rhodopsin function in photoreceptor cells (52). Inducible promoters also are useful for defining the function of particular cells, since the cells can be ablated by the induction of a gene encoding a toxin. This has proved useful in many studies designed to define cell function or signaling mechanisms (17,23,28,31,33).

An inducible promoter system that is widely used for tissue-specific expression in *Drosophila* is based on employing the GAL4 transcription factor of yeast (7). A large number of transgenic lines have been produced that express GAL4 under a specific *Drosophila* transcriptional enhancer. The GAL4 binding site is inserted upstream of a gene that one wishes to activate in a tissue-specific manner, and a transgenic line containing this gene is produced by *P* element-mediated transformation. This gene will only be expressed when the line is crossed to one of the lines expressing GAL4 in a tissue-specific pattern. This powerful technique is proving effective for ectopic expression (11), mutant rescue (26), and toxin-mediated ablation approaches (28).

P elements are useful tools in generating *Drosophila* null or hypomorphic mutants that define the function of a particular gene. Antisense RNA can be expressed from an inducible promoter to block gene expression (given the caveats of designing an appropriate antisense molecule and the possible failure to completely eliminate gene expression). A null phenotype can be achieved using ribozymes designed to recognize and cleave specific messenger RNAs (1,49,51). These constructs typically are regulated by an inducible promoter so that transgenic stocks can be maintained wherein deleterious phenotypes are observed only upon promoter induction.

A common strategy for making mutants using *P* elements is the mobilization of endogenous *P* elements through the use of a strain that constitutively expresses transposase (38). By mobilizing elements, genetically eliminating the transposase-encoding gene, and screening for mutants over a genetic deficiency or known mutant allele, new *P* element-induced mutations may be obtained. The *P* element and its surrounding target DNA can be identified by constructing a DNA library from the mutant and screening with a *P* element probe or by plasmid rescue (described below). In designing mutagenesis schemes, it is useful to note that proximity of a *P* element to a target site one wishes to mutate increases the chances of successful mutagenesis (13,48).

Chromosomal rearrangements can be induced using the *P* element system. When elements are located *in cis* on a chromosome, induction of transposase can result in deletion of the intervening region (8). Genomic rearrangements also can be induced through the use of the FLP recombinase system from yeast (13). When the FLP recombinase target (FRT) sites are inserted by *P* element-mediated transformation, the FRT sites can be excised by induced expression of FLP. Under these conditions, FRT sites on different chromosomes can recombine yielding chromosomal deficiencies and duplications.

The FLP/FRT system is also useful for the induced removal of engineered genes introduced by *P* element-mediated transformation (13,14). When the FLP recombinase is expressed under an inducible promoter, direct repeats of the FRT target that flank a transposed gene are removed along with the transposed gene. This allows one to study the effects of terminating expression of a particular transgene. In other types of constructs, gene expression may be induced by FLP-mediated removal of a *P* element insert within a gene, assuming the remaining *P* element termini permit gene expression.

A final application of *P* elements is to identify the location of transcriptional enhancers and their adjacent genes by enhancer trapping (6,35,50). A *P* element is engineered to include a weak promoter and a reporter gene such as *E. coli lacZ*. This enhancerless construct is only expressed when it is inserted near a transcriptional enhancer in the *Drosophila* genome. The expression pattern, as detected by observation of reporter gene expression, identifies the stage and tissue-specificity of enhancer function. By cloning the sequences flanking the inserted element, a gene expressed in an interesting pattern can be identified. This is most often accomplished by the plasmid rescue technique. Here, an *E. coli* origin of replication and an antibiotic resistance gene are included within the reporter gene construct. By cutting the *Drosophila* genomic DNA with a restriction enzyme at sites that flanks the *P* element, circularizing the resulting DNA using DNA ligase, transforming the DNA into *E. coli*, and obtaining the appropriate transformants by antibiotic selection, one can identify the *Drosophila* DNA located within or near to the enhancer and gene of interest. This can also be achieved by employing the polymerase chain reaction on the circularized DNA. Subsequently, one should be able to identify the gene of interest by obtaining flanking DNA through library screening or from the *Drosophila* genome project.

GENERAL CONSIDERATIONS

Mechanism of DNA Uptake and Integration

DNA is delivered to the *Drosophila* embryo by injection using a micropipet. Injected DNA consists of the DNA for insertion into the genome surrounded by the inverted repeats of the native *P* element and the helper plasmid encoding the *P* element transposase. Upon injection, the transposase gene is transcribed, and the transposase protein is synthesized within the developing embryo. Transposase is responsible for recognizing the ends of the DNA to be inserted into the genome and choosing the target site for insertion. There is a rough consensus sequence for insertion (34), and numerous studies have shown that *P* elements appear to prefer to insert in promoter regions of transcription units. Some of the molecular details of element insertion were discussed in the section entitled Historical Overview of *P* Element-Mediated Transformation.

General Procedure and Materials

Most available vectors for *Drosophila* transformation are based upon the *P* element, although other transposable elements can also be used (e.g., Reference 12). DNA constructs used for transformation contain four major components. *(i)* Terminal repeats from the *P* element, which are required for transposition into the chromosome. Any transforming plasmid contains these two sequences that flank the section of DNA to be inserted. Notably, the plasmid used is a nonautonomous element, meaning that it does not

Table 1. Examples of Marker–Strain Combinations Used for *P*-Element Transformation in *Drosophila*

Host Strain (Phenotype)	Selectable Gene	Selectable Phenotype	Vector Examples
w^{1118} (white eyes)	*white*$^+$	Rescue of white eye color to yellow, orange, or red	CaSpeR series, placW
ry^{506} (rosy eye color)	*ry*$^+$	Rescue of eye color from rosy to wild-type red	Carnegie series, p(PZ)
Any genotype	*neo*^{R}	Resistance to G418	pUChsneo

contain sequences encoding the transposase enzyme. It is therefore unable to induce its own transposition and, once integrated, is stable in the genome unless a source of transposase is subsequently introduced. *(ii)* A multiple cloning site (polylinker) in which to introduce the DNA of interest. The precise sequence of this site varies depending upon the type of vector used. *(iii)* A gene that serves as a positive selectable marker for transgenic animals. To identify the individuals that stably carry the introduced plasmid, a number of different selectable markers are available. It is important that the selectable marker carry a dominant allele of a gene, since transgenic individuals will be heterozygous when first isolated. The most common gene–marker combinations are summarized in Table 1. *(iv)* Plasmid sequences that permit growth and amplification of the DNA in *E. coli* cells. Since a relatively high concentration of DNA is required for injection, all vectors currently employed carry plasmid sequences that confer high copy number of the plasmid in bacterial cells. The structure of a typical transformation vector is shown in Figure 2.

A large number of vectors exist for *P* element-mediated germline transformation. These vary in both their selectable marker and in the cloning sites available for insertion of DNA. A comprehensive list is available on the World Wide Web at Flybase, the Drosophila genetic database (URL: **http://flybase.bio.indiana.edu:82/transposons/ lk/DIS.html**). This site contains links to the complete sequences of the vectors and has annotated maps describing their structures.

Fly strains used as recipients for DNA injection, and any strains used for subsequent crossing experiments, should be *M* strains that lack endogenous *P* elements. For transgenic flies, it is important that the inserted element remain stably inserted into the chromosome; any mobilization induced by endogenous *P* elements might result in loss of the element or further mutagenesis of the genome. The precise recipient strain to be used for injection depends upon the type of vector used. Recipient fly lines for specific constructs are detailed in Table 1. In general, if the marker is a wild-type version of an endogenous *Drosophila* gene, then one normally injects into embryos with a background mutant for that gene so as to be able to screen readily for transgenic progeny. Furthermore, it is usually preferred that the mutation is a small deletion rather than a point mutation, so that the probability of isolating revertants is reduced. As such, w^{1118} is frequently used as the recipient strain for vectors carrying w^+, and ry^{506} is used for vectors carrying ry^+, since both mutant alleles arise from deletions (29). If the selectable marker confers drug resistance, such as G418 resistance conferred by neo^R, most strains can be utilized as recipients.

The success of stable transformation in *Drosophila* depends upon the introduction of the transforming plasmid into the cells of the presumptive germline. Since the embryo develops as a syncitium during its first two hours

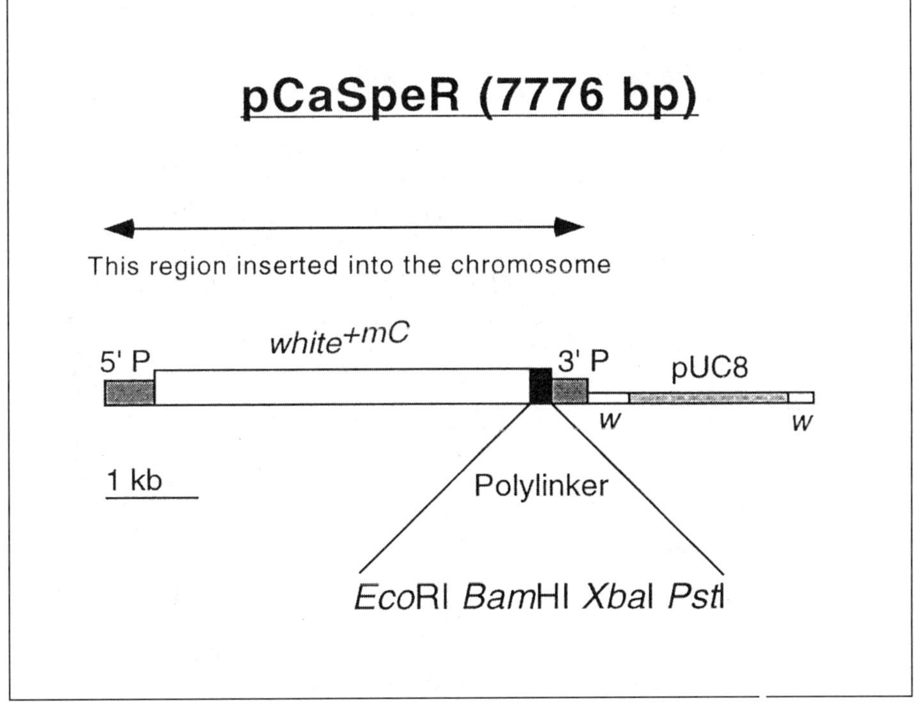

Figure 2. Organization of the pCaSpeR *P*-element transformation vector. 5′ *P* and 3′ *P* refer to the *P*-element termini that flank the DNA to be inserted into the chromosome; w^{+mC} is a mini-white gene used as a marker for introduction of the element into *Drosophila*; unique restriction sites in the polylinker are indicated. The plasmid also contains pUC8 sequences, which permit DNA amplification in *E. coli* with ampicillin selection. *w* refers to small fragments of the white gene that are also present in the plasmid.

(at 25°C), where the fertilized egg undergoes 13 cycles of nuclear division without concurrent cellular division, DNA introduced at an early stage is accessible to all nuclei. Thus, localization of the DNA in the cell does not require sophisticated intracellular injection techniques.

The entire germline of both male and female *Drosophila* derives from the pole cells, a group of cells that form at the posterior tip of the embryo. These are the first cells to cellularize after nuclear cycle 13, and they subsequently migrate dorsally and anteriorly during embryogenesis. DNA is injected at the posterior tip of the embryo so that it may access the presumptive germline.

The outcome of the transposition of foreign DNA into *Drosophila* embryos is an adult with a chimeric germline, with some cells genetically unchanged and others containing the modified *P* element in their chromosomal DNA. It is important to appreciate that this adult usually will be phenotypically unchanged in regard to the screening marker, since only a small proportion of its cells will carry the element. For example, a w^{1118} fly that was injected as an embryo with a construct derived from CaSpeR, and therefore carrying w^+, will still have white eyes. However, this fly may give rise to some offspring that carry a copy of the transposed dominant w^+ gene in all tissues, conferring a red eye color.

Some nomenclature rules have been developed that are specific to the *P* element transformation procedure. Organisms injected with the DNA as embryos are termed the G0 generation (41). Offspring of these individuals are called the G1 generation. It is the G1 generation that is screened for isolating transgenic individuals. Tranformants are usually backcrossed to yield a G2 generation from which to establish individual lines. To name transgenic lines, Ashburner (3) suggests the designation *P[a b] c* where *P* indicates a *P* element-derived transgenic line, *a* refers to the selectable marker (for example w^+), *b* refers to the specific gene or construct that is being studied, and *c* refers to the designation of the independent line carrying the construct.

General Description of the Procedure and Variations

The procedure of *P* element-mediated germline transformation can be divided into three major sections: *(i)* isolating and preparing embryos for injection; *(ii)* the injection process; and *(iii)* postinjection care and crossing to isolate stably transformed lines. To process the embryos in an efficient manner, the first two are performed over a 1-hour period.

To prepare harvested embryos for injection, the eggshell, or chorion, is removed so that the injection needle will easily penetrate the vitelline membrane surrounding the embryo. Next, the embryos are arranged on an agar plate so that they lie side-by-side and are oriented in the same anterior-posterior direction. Then embryos are lifted from the plate by adhering them to a slide carrying a strip of double-stick tape. Next, the embryos are desiccated slightly to accommodate the increase in embryo volume resulting from injection of the DNA solution. Finally, embryos are covered with oil to prevent further desiccation.

Prepared embryos are injected with the DNA solution using a micropipet mounted on micromanipulators and observed using an inverted microscope. The slide carrying the desiccated, oil-coated embryos is placed under the microscope such that the posterior ends of the embryos are facing toward the needle. Injection at the region of the forming pole cells is accomplished by puncturing the posterior of the embryos with the needle and depositing a small amount of DNA-containing solution near the pole. Once one embryo has been injected, the needle is removed from the embryo and the slide control of the microscope is used to move the adjacent embryo into view for injection.

Following injection, the embryos are incubated at 25°C until hatching, when they are removed to a vial containing fly food and allowed to develop to adulthood. Since the aim is to generate transgenic germline cells, these G0 adults will generally show no signs of the marker phenotype (an exception is the *ry*⁺ marker, where transient expression of the unintegrated plasmid can produce sufficient xanthine dehydrogenase activity to rescue the eye color). Following eclosion of the G0 adults, each is crossed independently to flies of the opposite sex (the strain of flies used to cross will depend upon the marker being used), and the G1 offspring are screened for individuals carrying the appropriate marker.

Given the complicated procedures involved in preparing embryos and isolating transgenic individuals, there are few alternatives to the approach described. Kamdar et al. (18) described a method for transient introduction of DNA into embryos via electroporation, however this approach has not been applied to *P* element transformation. All fly laboratories generating transgenic animals utilize some form of the general procedure described in this chapter. Some minor variations are described in the section entitled Alternative Approaches.

PROTOCOLS

Major Equipment

Dissecting Microscope and Light Source

A dissection microscope is required to align the harvested embryos prior to injection, for collecting hatched larvae from the slide, and for identifying marked flies for genetic manipulation. We prefer to use a light source with fiber-optic guides to reduce heating of the organisms, particularly during embryo orientation. High-quality optics are not required for these steps.

Needle Puller

A machine to generate a needle with a sharp taper and fine tip is required. A reasonable option is the PUL-100 from World Precision Instruments (Sarasota, FL, USA). Alternatively, needles can be purchased from the same source.

Inverted Microscope for Viewing Injection Procedure

A microscope with 200× bright field magnification is required to view and coordinate the injection process. We use an inverted microscope that provides a large working distance for the needle holder and micromanipulator. Suitable models include the Nikon TMS (Nikon, Melville, NY, USA). Some laboratories utilize a compound microscope instead, however the working distance is relatively small.

Micromanipulator for Positioning Needle

Fine control of the needle is essential to prevent unnecessary damage to the embryos. Commonly used systems provide fine and coarse control in the x, y, and z axes, to allow injection of embryos on slides in a fixed-level stage. In other systems, the position of the needle is fixed in two dimensions and will only move along the z axis; x-y motion (including puncturing the embryos with the needle) is provided by movement of the microscope stage.

Syringe Setup for Injection

The flow of the DNA-containing solution through the small bore of the injection needle is quite slow and requires a significant amount of force. To achieve this, we attach the needle to Teflon® tubing, which is in turn connected to a 50-mL plastic syringe via a metal luer lock. Air pressure generated from depressing the syringe plunger is sufficient to expel liquid from the needle.

Detailed Protocols

An overview of the entire transformation procedure is given in Table 2 and is described in the sections that follow. This starts 2 weeks before the actual injection date, when bottles of fly food need to be inoculated to provide adults for the fly cages. This is followed by the steps carried out on the day of injection, the postinjection care of the embryos, and subsequent fly crosses aimed at isolating transformed individuals. The procedure is completed 3 to 4 weeks after the injection date with the isolation of transformed individuals (the G1 generation).

Table 2. Overview of Major Steps in *P*-Element Transformation

Day[a]	Steps to Perform	Sections in Text
-14	Set up ten bottles of host flies (to be used for injection).	Preparation of Flies and Cages—Temperatures and Incubators
-2	Transfer approximately 500 young adults to fly cage.	Preparation of Flies and Cages—Temperatures and Incubators
0	Harvest and inject embryos. Set up 10 bottles of host flies (to be used for crosses on Day +9 to +15).	Embryo Collection and Alignment through Embryo Injection Table 3
+1, +2	Collect hatched larvae and transfer to yeast-containing vials.	Embryo Incubation and Larvae Collection
+9 to +15	Collect males and virgin females from the bottles set up on Day 0.	
+10 to +15	Collect G0 adults and cross to the flies isolated above.	Identifying Transformants
+21 to +30	Isolate transformants in the G1 generation.	Identifying Transformants

[a]Day 0 is the day of injection.

Preparation of Flies and Cages

<u>Temperatures and Incubators</u>

A large population of healthy flies is essential to obtaining sufficient embryos for injection. During periods of injection, we maintain 20 bottles of flies of the appropriate genotype in incubators at 25°C; we inoculate 10 new bottles each week and discard those that are 2 weeks old. These bottles will be used as a source of flies both for setting up cages and for crossing G0 adults. If growth is poor in a particular bottle, it is discarded.

Bottles should to be set up 2 weeks before the anticipated injection date so that sufficient numbers of flies are available for populating cages. Two to three days before injection, approximately 500 young adults are transferred from the culture bottles to the fly cage; this usually corresponds to the contents of 2 bottles.

The fly cage is a 500-mL plastic tri-pour beaker, over the end of which is affixed a 90-mm petri plate containing grape-juice agar (see the section entitled Support Protocols). The plate has a smear of yeast/water paste on it to serve as food. Holes are punctured in the side of the beaker to prevent excessive moisture build-up, and the petri plate should be replaced at least once a day so that larvae do not hatch inside the cage (Figure 3A).

The cages should be placed in a dark place out of normal traffic. Appropriate options are a closed 25°C incubator or a drawer at room temperature that should not be opened. Gentle treatment of the flies serves two purposes: *(i)* undisturbed flies tend to lay more eggs; and *(ii)* disturbed flies retain fertilized eggs rather than laying them, which can result in harvesting many eggs that are too old for injection. Figure 3B shows a petri plate upon which a number of eggs have been laid around the smear of yeast paste.

<u>Embryo Collection and Alignment</u>

The procedures in this section and the section entitled Embryo Injection are summarized in Table 3, which can be used as a guide to the major steps carried out on the injection day. Appropriate times are indicated.

<u>Purging Flies of Retained Embryos</u>

To encourage the caged flies to void any fertilized eggs they are carrying and that may be too old for injection, the eggs from the first hour of collection are discarded.

<u>Collection Period</u>

Although there can be some variability in the length of an egg collection

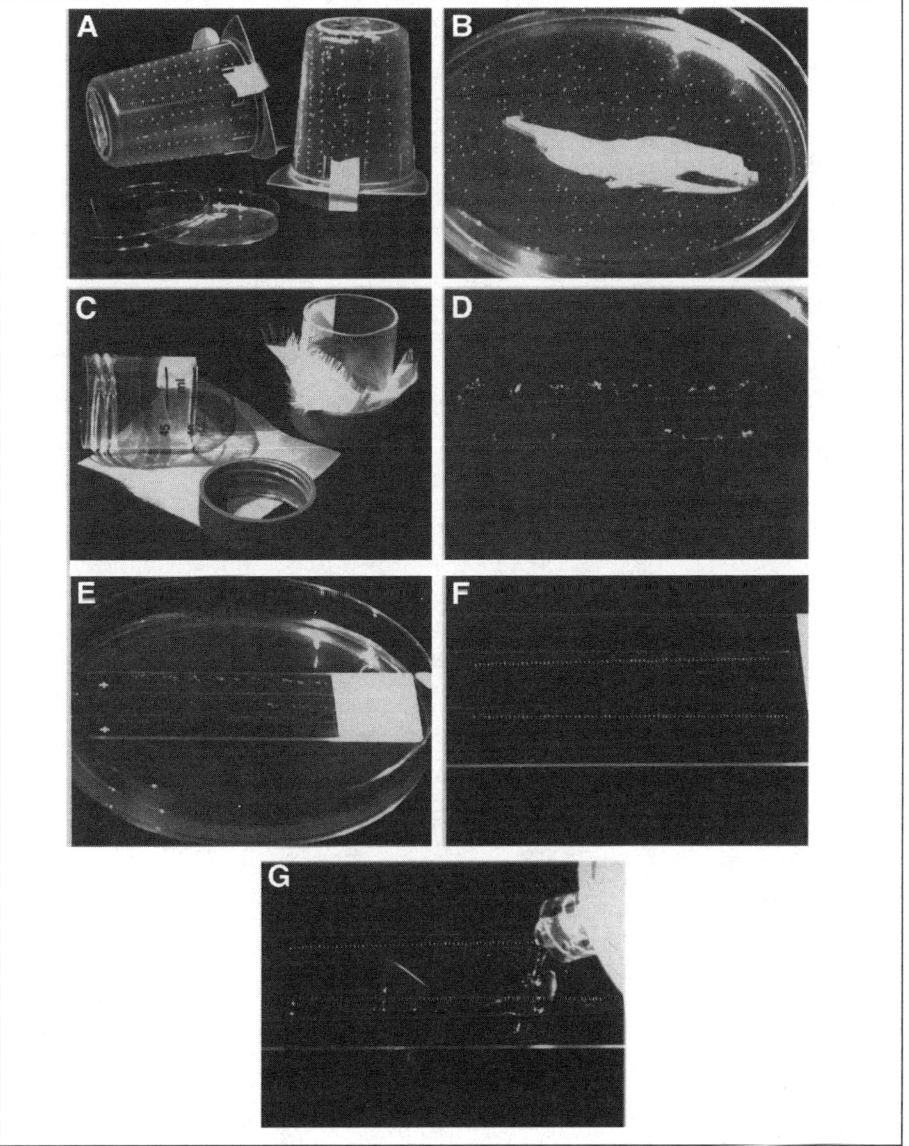

Figure 3. Steps taken to ready embryos for injection. (A) Fly cages are prepared from a plastic beaker with holes punctured in it, over the end of which is affixed a petri plate with grape/agar food. (B) Flies lay eggs on the petri plate around a smear of yeast paste that is used as a food source for the flies. (C) Eggs are rinsed from the plate into a sieve, made from a cutoff 50-mL plastic centrifuge tube and nylon mesh. (D) After dechorionation, embryos are arranged along lines scored into a grape/agar plate; shown are two lines of oriented embryos as well as clumps of unused embryos. (E) To transfer the embryos from the plate to a microscope slide, a slide with strips of double-stick tape is laid over the embryos on the plate and depressed slightly to affix the embryos to the tape. (F) The slide is then lifted from the plate with the embryos affixed to it. (G) After a period of desiccation, the embryos are overlaid with oil to prevent further drying. The embryos are now ready to inject.

Table 3. Overview of Major Steps on the Day of Injection

Time of day[a]	Steps to Perform	Sections in Text
08:00	Place a fresh plate over the cage of flies.	Purging Flies of Retained Embryos
09:00	Remove and discard the 08:00 plate; place a fresh plate over the cage.	Purging Flies of Retained Embryos and Collection Period
10:00	Remove and keep the 09:00 plate; place a fresh plate over the cage.	Purging Flies of Retained Embryos and Collection Period
10:00-10:03	Harvest embryos from plate into sieve.	Harvesting Embryos
10:03-10:06	Dechorionate by immersion in bleach.	Dechorionation and Washing
10:06-10:08	Wash with water, then blot dry.	Dechorionation and Washing
10:08-10:11	Transfer embryos from sieve to scored agar plate.	Orientation of Embryos En Masse
10:11-10:22	Orient embryos on plate.	Orientation of Embryos En Masse
10:22-10:23	Count embryos.	Orientation of Embryos En Masse
10:23-10:25	Adhere embryos to slide carrying double-stick tape.	Picking Up Embryos with Sticky-Tape Slide
10:25-10:26	Place slide in dessication box.	Embryo Desiccation
10:26-10:36[b]	Dessicate.	Embryo Desiccation
10:36-11:00	Cover embryos with oil, then inject embryos.	Immersion in Oil and Embryo Injection
11:00	Remove and process the 10:00 plate; place a fresh plate over the cage; start new injection series.	

[a]The start time is a suggestion only. Late risers can accommodate as they see fit.

[b]Times for dessication will need to be empirically determined, therefore the times shown are only intended as a guideline. However, dessication and injection will ideally be completed before 11:00, such that the next injection series can be started on time.

period, we collect embryos for injection for 1 hour. Shorter collection times may not allow the flies to settle and lay eggs. Longer times may produce some embryos that are too old for injection once they have been processed. Ideally, the steps, in the Harvesting Embryos procedure through the Where and How Much to Inject procedure below, should be completed within 1 hour, so that the next plate of embryos can be processed immediately after the first set has been completed.

Harvesting Embryos

Embryos are harvested from the collection plate by rinsing with a solution of NaCl/Triton® X-100 (see the section entitled Embryo Collection Solution) and rubbing gently to release them from the surface of the plate. This embryo suspension is then poured through a small sieve to separate the embryos from the wash solution. Sieves are made from 50-mL polypropylene centrifuge tubes with screw lids. The barrel of the tube is cut to leave approximately 50 mm of tube attached to the cap, and the bottom portion of the tube is discarded. The safest way to cut these tubes is to use a pipe cutter obtained from a hardware store. The cap of the tube is removed, and the center is cut away to leave the outside ring of thread. The cap is then screwed back onto the tube barrel after sandwiching a small piece of mesh between the two plastic components (Figure 3C).

Dechorionation and Washing

Embryos are dechorionated by immersing the end of the sieve containing the embryos in 50% (vol/vol) bleach for 3 minutes. During immersion, the sieve should be swirled to break up clumps of embryos. After removal of the eggshell, traces of bleach are eliminated by washing the embryos in the sieve with copious amounts of water. Sometimes embryos will stick to the sides of the sieve since they are now strongly hydrophobic. These should be washed down to the mesh using a squirt bottle of water. The embryos are blotted dry by touching tissue paper to the mesh below them.

Orientation of Embryos En Masse

Embryos from the sieve are transferred to a grape/agar plate. The middle of the plate should first be scored with a line using the side of a glass microscope slide. Embryos are then arranged along this line so that they are not touching each other, and such that their anterior–posterior axis is perpendicular to the line. At this stage, the embryos should be slightly damp but not wet. For ease of injection, embryos should be arranged so that they face the same direction, e.g., with anterior toward the line. Embryos arranged on a plate are shown in Figure 3D. The anterior and posterior ends of an embryo may be distinguished by the presence of a small protuberance called the micropyle at the anterior end (see Figure 4C). With experience, the ends may also be distinguished by their shape: the posterior end is more blunt and rounded than the anterior. If embryos have an opaque rather than shiny appearance and two long filaments protruding from the anterior end, these should be avoided. These are embryos from which the chorion was not removed.

In order to maintain the injection schedule, we orient as many embryos as possible during a 10-minute period. In this way, we are able to comfortably inject all the embryos that have been lined up before the start of the next injection series. With experience, one person can line up and inject up to 200 embryos per hour. For this purpose, we use a grape/agar plate scored with two parallel lines such that two rows of embryos are arranged (see Figure 3). After the 10-minute aligning time is over, the number of lined-up embryos should be counted and recorded. Any nonaligned embryos should be removed from their vicinity, so that they will not be carried over to the next step.

Picking Up Embryos with Sticky-Tape Slide

To transfer the embryos from the scored plate to a microscope slide, a slide is prepared with a strip of double-stick tape along its length. This slide is then laid over the plate so that the tape comes into contact with the line of embryos (Figure 3E). The embryos are cushioned from damage by the soft agar. The slide should be depressed slightly to ensure that all of the embryos adhere to the tape. The slide is then lifted from the plate, labeled, and is ready for desiccation (Figure 3F).

Embryo Desiccation

Embryos are desiccated by placing the slide, embryos facing up, in a box containing desiccant such as Drierite® (Aldrich, Milwaukee, WI, USA). The desiccant should contain an indicator so that its usefulness can be assessed, although the grade size of the fragments is not important. The box should have a tight seal so that the desiccant does not become rapidly exhausted, and so that consecutive desiccation periods will be consistent.

The length of the desiccation time can vary from 5 minutes to as many as 20 minutes in humid climates, and therefore can only be empirically determined by injection (see below). A number of factors can contribute to optimal desiccation time. These include ambient temperature and humidity, the dampness of the embryos during orientation, the length of the orientation time, and the efficacy of the desiccant. Also, manually dechorionated embryos (see the

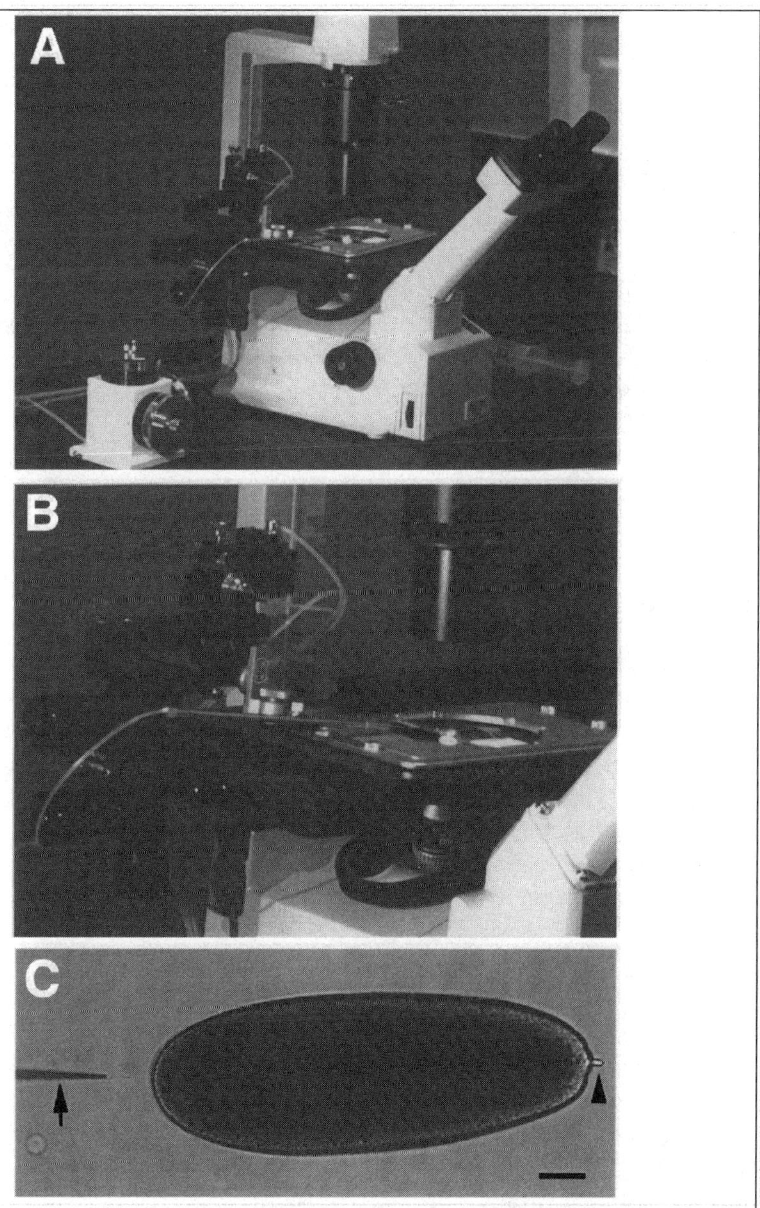

Figure 4. Arrangement of the microscope slide relative to the injection needle. (A) Overview of injection setup, showing the inverted microscope with (to the left of the stage) the injection apparatus. The manipulators are affixed to the left side of the stage; the Narishige MM203 (Narshige International USA, East Meadow, NY, USA) control (far left) produces fine movements of the needle via hydraulic lines. (B) Close-up of needle holder relative to the microscope slide and stage. In this arrangement, the needle approaches the slide from the left. Note that the axis of the needle is perpendicular to the axis of the slide (see text). (C) A dechorionated embryo with the needle (arrow) close to the posterior end. Note the micropyle (arrowhead) which demarcates the anterior end. Bar, 100 μm.

section entitled Manual Dechorionation of Embryos) require a shorter desiccation. For all these reasons, consistency in timing and consistency in actions is essential, so that logical adjustments can be made to optimize conditions.

Immersion in Oil

After desiccation, the slide is removed from the box, and the embryos are covered with a layer of oil to prevent further drying (Figure 3G). Halocarbon Series 700 oil (Labscientific, Livingston, NJ, USA) must be used for this, since it is one of the few oils that is nontoxic to the embryos, and it allows gaseous exchange. This oil is quite viscous, so that it does not run off the embryos and expose them to the air. However, too much oil can decrease oxygen availability and reduce viability of the embryos; too little oil will expose them to drying out in the air. The embryos are now ready to be injected.

Embryo Injection

Loading DNA Solution into Needles

The most important step in loading the needle is to centrifuge the DNA mixture for 3 minutes before loading. This removes particulate matter that will inevitably clog the needle. This step is particularly important since many DNA purification procedures employ resin columns, and a small amount of particulate matter often is recovered in the eluate. Needles that have been prepared as described in the section entitled Pulling Needles are generally closed at the tip and so must be backfilled with the DNA mixture. Capillaries used for making needles have a fine groove (filament) along the inside edge, facilitating DNA uptake by capillary action. There are several simple procedures for introducing DNA into the capillary. One is to use a pulled-out glass micropipet to physically squirt the DNA inside the barrel of the needle; another is to stand the needle, point upwards, in a microcentrifuge tube containing 10 µL of DNA solution and to allow the needle to fill via capillary action. A final method is to hold the needle vertically, tip downwards, to slowly pipet 1 to 2 µL of solution onto the open end of the needle and to allow the DNA to enter the needle over the course of approximately 5 minutes. Needles can be loaded with DNA in advance, although we generally load during the desiccation period and then reload whenever a fresh needle is desired. Needles loaded too long in advance are in danger of having the DNA solution dry out in them.

Breaking the Needle Tip

The sealed end of the needle must be broken off before it can be used for injection. Equipment exists for grinding the tip of the needle to produce a sharp beveled tip (see the section entitled Alternative Approaches). An alternative approach is to mount the needle onto the manipulators and, while viewing under the inverted microscope lens, gently tap the tip of the needle against the side of a microscope slide. If a small amount of air pressure is maintained inside the needle by pressure upon the syringe, a pool of DNA will flow onto the slide, indicating that it has been broken successfully. Caution should be taken to ensure that only just enough force is used in breaking the needle as is necessary; the smaller the hole in the tip of the needle the more successful will be the injection.

<u>Where and How Much to Inject</u>

The slide should be placed on the microscope stage such that the posterior ends of the embryos are pointing toward the needle, either to the left or the right (Figure 4). Since the embryos are mounted on the slide perpendicular to the long axis of the slide, the slide holder of the microscope should hold the slide appropriately. This is important, since not all microscopes accommodate that orientation of the slide, and a slide holder might have to be custom made.

It will take time and experience to assess the status of the embryos. The ideal embryo is relatively rigid, permitting easy puncturing with the needle but no loss of cytoplasm upon needle removal. Two suggestions are helpful in this regard. The first is accurate record keeping, so that the researcher can monitor the proportion of injected embryos that subsequently hatch. This proportion should approach 75% in ideal situations. The second is that the efficacy of desiccation times should be assessed only when a significant proportion of the embryos from a slide have been injected. This ensures adequate numbers of embryos are being assessed to allow one to draw conclusions.

One aspect of successful injection is the rapid insertion of the needle into the embryo, without distending the embryo too much in the process. This is accomplished best using a fine needle, particularly one broken so that the tip is beveled. Accurate puncturing of the embryo is also important, and this is achieved by ensuring that the tip of the needle is perpendicular to the point of contact with the embryo. For this purpose, the needle may need to be manipulated in the z axis; thus, fine control in the vertical direction is an important requirement for injection manipulators. An injection in progress is shown in Figure 5A.

Once the needle has penetrated the embryo, a small amount of DNA solution should flow into the embryo. This may be encouraged by gentle pressure on the syringe barrel. Alternatively, if there is a slow continuous flow from the needle, no additional pressure is required. The appropriate volume of solution to inject is difficult to judge since there is no calibration on the needle itself.

However, the observer will notice swirling of the cytoplasm at the tip of the needle during injection; a useful rule of thumb is that if swirling is seen, then approximately the correct amount of DNA solution has been introduced.

Upon removal of the needle from the embryo, there should be no loss of cytoplasm from the embryo. An example of such leakage is shown in Figure 5B. The major cause of cytoplasm loss is insufficient embryo desiccation; if cytoplasmic leakage is observed, the desiccation time should be increased for the next injection series. Another factor that can contribute to cytoplasm loss is the size of the needle; a large bore needle will more likely cause leakage than a small bore needle. Embryos can survive some loss of cytoplasm, but losing too much will be lethal. Although moderate cytoplasm loss will result in survival of adults, these adults may be sterile since the lost pole plasm is involved in determining the germline. It is also possible to over-desiccate the embryos; if the embryos easily distend when prodded with the needle (Figure 5C), they are over-desiccated and will not survive embryogenesis.

Once all of the embryos have been injected on one slide, the slide is moved to a humid container within a 25°C incubator to allow the embryos to develop and hatch. Care should be taken to ensure the slide is flat, so that the oil remains in place over the embryos. In instances where the same needle is going to be used for the next slide of embryos, we leave the slide on the microscope stage with the needle immersed in the oil until the next slide is ready. This prevents the DNA solution from drying out at the tip of the needle.

Embryo Incubation and Larvae Collection

Following injection, the embryos are allowed to develop and hatch before being placed into vials of fly food. At 25°C, it will take 24 to 48 hours for all of the embryos to hatch. Larvae swim around in the oil in which they were immersed. At this time, they are scooped from the slide with a paintbrush and placed into fly food vials in groups of 50. It is important that they be close to the medium in the vial so that they do not die before finding the food. A few grains of active dried yeast are sprinkled on the food so that the larvae will have living yeast to eat as they grow. The vials are incubated at 25°C until the G0 larvae develop into adulthood; this will take approximately 10 days.

Since the G0 adults will be crossed to other adults to screen for transformed individuals, fresh bottles of stock flies should be set up on the day of injection, so that newly eclosed G0 adults can be mated to freshly collected male and virgin female flies.

Identifying Transformants

After eclosion, all of the G0 flies are separated and each is crossed to 3 to 5

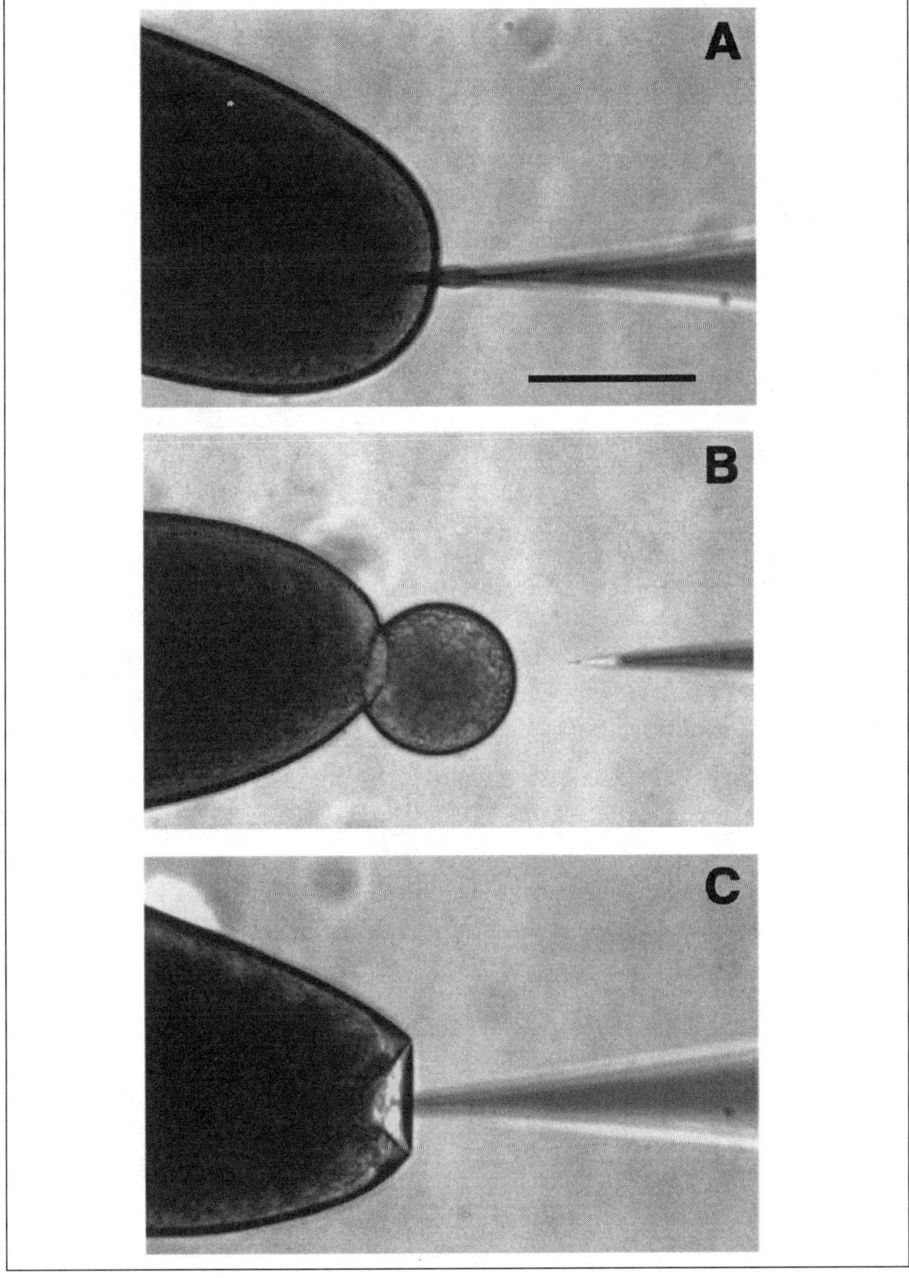

Figure 5. Steps during the injection process. (A) The needle is inserted through the vitelline membrane, and a small amount of DNA is deposited at the posterior end. The needle is then rapidly withdrawn. (B) If the embryo has not been properly desiccated, cytoplasm leaks after withdrawal of the needle. This will result in either sterility or lethality. (C) If the embryo is over desiccated, it loses its rigidity and is difficult to puncture. Over-desiccation will result in lethality. Bar, 100 μm.

flies of the opposite sex, in order to sample as many sex cells as possible. The offspring of these crosses (the G1 generation) will usually be the first to show the phenotype for screening. Such phenotypes are shown in Table 1. In this scheme, mating between G0 adults is avoided. This is because a cross of G0 parents that generates transgenic offspring may produce progeny containing transgenes donated by both parents. Although independent lines might be distinguished by genetic mapping or by Southern blotting analysis, the analysis is simpler if the G0 adults are separated before mating can occur. Therefore, it is important to keep a detailed record of when particular flies eclose, if males and females eclose at the same time, and if it is possible that any of the males and females might have mated.

An important aspect of studying transgenic lines in *Drosophila* is the need to isolate several independent transgenic lines for any particular construct, obviating any problems associated with position effects resulting from transgene insertion site. One G0 fly may give rise to one transgenic offspring; alternatively, up to 50% of the offspring may be transgenic. This reflects the time during germline cell development at which the *P* element inserted into the chromosome, since proliferation of the cells occurs during development. Therefore, multiple transgenic offspring from the same G0 parent may correspond to the same original transposition event and would therefore be non-independent lines. However, some of the transgenic offspring may represent independent insertion events that occurred in the germline. In this case, more than one independent line may be isolated from a single G0. However, it will be relatively time-consuming to determine this using Southern blotting or genetic mapping.

Mapping Chromosomal Locations of Inserts

<u>Confirming that Lines are Independent</u>

If a transformant (a G1 individual showing the marker phenotype) is the offspring of a G0 individual that was separated from other G0 adults prior to mating, this represents an independent transgenic line. If G0 adults were not separated prior to mating, or if multiple transformants are produced by the same adult, it may be necessary to determine if they represent independent insertion events. This can be accomplished by determining if the inserted DNAs are located in different regions of the genome. Genetic crosses using marked balancer chromosomes can indicate which chromosome the transposon has inserted into. Lines arising from insertions on separate chromosomes are independent; however if two lines map to the same chromosome they may or may not be independent. Fine-detail mapping to distinguish separate insertion events may be achieved by Southern blotting of chromosomal

DNA. The principle of this is to digest the chromosomal DNA with an enzyme that cuts once inside the transposon and also at a chromosomal site flanking the insert. The distance between the internal site and the external site will most likely be different for different lines. Southern blotting will reveal bands of different sizes in independent lines and identical sizes in identical lines.

Productivity Considerations

Timetable for Procedures

The timetables provided in Tables 2 and 3 are designed to achieve the maximum productivity in the time allowed. Within each hour, approximately 30 minutes is dedicated to preparing the embryos for injection, and the remaining time is required to inject prepared embryos. Once this has been completed, a fresh set of 0 to 1-hour-old embryos should be available in the fly cage. If these times are adhered to, one may continue injecting indefinitely on the same schedule. As long as the grape/agar plates are changed on time, it is also easy to break for 1 hour at any time during the day.

One Versus Two People

Since each component of the injection hour takes approximately 30 minutes, we have been able to double the productivity by having two people work on the injections at the same time, with no requirement for additional equipment. For example, person one would start preparing embryos at 10:00. These would be ready to inject at 10:30, and injection would be completed by 11:00. Person two would start preparing embryos from a separate population cage at 10:30 (after person one has finished using the dissection microscope) and would start injecting at 11:00 (when person one is starting their second injection series of the day). A similar alternative would be to have person one continuously preparing embryos (first at 10:00, next at 10:30) and to have person two continuously injecting (first at 10:30 then at 11:00).

Support Protocols

Preparing Collection Plates

Prepare grape/agar plates by mixing 150 mL of grape juice with 450 mL water and boiling with 12 g of agar in a microwave oven. Once the mixture has boiled thoroughly, allow it to sit until the container is just cool enough to touch comfortably and then pour into plastic 90-mm petri plates.

Preparing Sticky Tape Slides

The simplicity of making microscope slides with double-stick tape belies the importance of this step. Before adding tape, the slides should be washed with ethanol to remove any residuum that might be toxic to the embryos. Lay lengths of double-stick tape over the slides and allow them to settle in place; bubbles of air between the slide and the tape should be avoided to enhance the optics under the microscope. To reduce the total amount of tape on the slide, we generally pare it down on each side until there is only a thin strip (approximately 3 mm) of tape present.

Most brands of double-stick tape contain a component that is toxic to the embryos. It is therefore essential that a reliable source of tape be used. Currently, Scotch Double-Coated Tape, catalog number 665 (1/2 in × 36 yd), is suitable. However, caution should be used when employing new batches in case the formulation has changed.

Preparing DNA for Injection

Successful transformation requires that the DNA be highly purified. DNA purified by CsCl gradient centrifugation is suitable as is DNA produced by more rapid column purification systems. We use DNA prepared with the Maxiprep system (Qiagen, Valencia, CA, USA) with great success.

DNA for injection is prepared as follows. In a microcentrifuge tube add 17.5 µg of transforming plasmid and 2.5 µg of helper plasmid. Add a tenth volume of 3 M sodium acetate, pH 5.2; mix, then add two volumes of 100% ethanol. Mix and centrifuge immediately for 5 minutes. Wash with 80% ethanol, dry the pellet, then resuspend in DNA injection buffer (see the section entitled DNA Injection Buffer). Before loading the needle with this DNA solution, the solution should be centrifuged for 3 minutes to pellet any particulate matter that might otherwise clog the needle.

Pulling Needles

Needles for injection are made by melting glass capillary tubes and pulling them out to a fine taper. Most capillaries used for this purpose have a small filament along the inside of the tube that aids in DNA uptake. Needles for injection should have a sharp taper so that they are not too flexible, but should have a fine tip so that minimal damage is caused to the embryo.

Table of Major Steps in the Procedure

Timetables of major events in the injection procedure are given in Tables 2 and 3. Table 2 covers the period commencing 2 weeks before the injection day to the time of identifying transformants in the G1 generation, some 3 weeks after injection. Table 3 describes a condensed timetable of the major events that take place on the day of injection. Each of these major events is cross-referenced between the table and the appropriate section of text (see the section entitled Detailed Protocols).

Alternative Approaches

Grinding Needle Tips

Instead of manually breaking the sealed tip of the needle to allow DNA to flow, one may use a micropipet beveler (for example the 1300M Beveler from World Precision Instruments). These items consist of a rotating abrasive disc onto which the needle tip is gently lowered until the sealed end is ground away.

Manual Dechorionation of Embryos

Some researchers prefer to manually dechorionate embryos rather than use bleach. This is achieved by placing the egg on a piece of double-stick tape on a slide, and then gently prodding and rolling it to break the eggshell. The embryo, enclosed in the vitelline membrane, can then be transferred to a fresh piece of tape for desiccation. Desiccation times for manually dechorionated embryos are significantly shorter.

Injecting through the Chorion

Robertson et al. (38) described a procedure to inject DNA into embryos directly though the chorion. This does not involve dechorionation or desiccation; DNA of a higher concentration than normal (0.5 mg/mL) is injected in small volumes by directly piercing the embryo through the eggshell. Although this represents a relatively straightforward procedure, it has not achieved widespread use, perhaps due to unreliability in results.

Alternative Strain and Vector Considerations

Although we routinely use transformation vectors carrying w^+ and inject into white-eyed lines, ry^+ vectors and neo^R vectors are still commonly used. These can be advantageous when attempting to use genetic crosses to bring together more than one transgenic construct in the same genome. Thus, one can select for rescue of eye color and survival on G418 in one simple step. One advantage of using w^+ is that homozygotes for a particular insert usually can be distinguished based upon a stronger red eye color than the heterozygotes; therefore, mapping is not necessarily required to maintain a homozygous stock.

Injection Machines

As an alternative to using a simple air-syringe setup for generating pressure to force the DNA out of the needle, automated injectors can be purchased that use compressed gas to deliver specific small volumes of DNA. Such instruments include the Eppendorf Transjector 5246 and a number of models available from World Precision Instruments.

EXPECTED RESULTS AND TROUBLESHOOTING

Numbers of Organisms Necessary for Success

Numbers of Embryos to Inject

Up to 200 embryos can be injected in a 1-hour period. The total number that should be injected for each construct will depend upon the proportion of those that will ultimately hatch. We inject between 200 and 600 embryos for each standard-sized construct, translating into a 1 to 3 hour injection period per construct.

Expected Percentage of Larvae Hatching

At least 30%, and as many as 75%, of the injected embryos should hatch into larvae. A general rule is to inject embryos and harvest hatched larvae until there are approximately 200 larvae.

Expected Percentage of Adult Survivors

Between 30% and 60% of the hatched larvae will emerge as adults. Of these, around 10% may be sterile because of damage to the pole plasm during injection. In general, we aim to isolate 100 G0 adults.

Expected Numbers of Independent Lines

A general rule is that at least three independent lines of any particular construct should be analyzed to ensure that the phenotypes that are seen are due to the construct itself rather than to effects caused by the insertion event. Ideally, more lines than this will be isolated in case of loss. The proportion of G0 flies that give rise to G1 transformants can vary a great deal, but should be at least 5% and may be as high as 30%.

Problems and Solutions

Problems with viability of the G0 animals are the major cause of reduced success with *P* element transformation. Some of these difficulties are genotype-specific. Since most laboratory strains of *Drosophila* are extensively inbred and may lack hybrid vigor, reduced viability can result from the vigorous treatment of the embryos during the injection process. Furthermore, the fecundity of some strains is poor, making it difficult to collect enough embryos in the 1-hour collection period. These types of problems are frequently insolvable, and it is simpler to inject the construct into a different strain. Other factors that cause reduced success can be addressed by the researcher, as described below.

Increasing the Number of Embryos Collected

Several factors affect the number of eggs that are laid by the flies in the cage. If the temperatures are extreme (outside of the 22° to 26°C range), this can reduce egg-laying. If the flies are disturbed, or if they are not kept in a dark environment, this can result either in reduced egg-laying or in the retention of eggs so that many of the embryos are too old to inject by the time they have been processed. Also, older flies (greater that a week old) or very young flies (less than 2 days) tend to lay eggs at reduced frequencies.

Increasing the Percent of Larvae Hatching

The number of embryos hatching is often the best short-term indication of eventual success at injection, and many factors contribute to this. Reduced viability can result from over-desiccation, under-desiccation leading to loss of cytoplasm during injection, toxicity from the double-stick tape, toxicity from contaminants in the DNA injection buffer, retarded development caused by the layer of oil being too thick, and desiccation caused by the layer of oil being too thin.

Most of these problems can be solved by simple controlled experiments. For example, toxicity associated with injection can be assessed by harvesting embryos, desiccating them, and covering them with oil, but not injecting them. Complications can arise when there are more than two causes of toxicity, but this may also be addressed experimentally.

Difficulties in Mapping Chromosomal Insertion Sites

Most mapping procedures designed to identify the chromosomal linkage of an insert are aimed at determining if the insert is on the X, second, or third chromosome by following the segregation of the transgene marker relative to markers on other chromosomes. If the insert does not fall into a simple linkage group based upon these experiments, then it may be on the small fourth chromosome or the Y chromosome; these are relatively rare occurrences. Alternatively, there may be more than one insert present in the line, making it possible to isolate the inserts and maintain two independent lines from one original G0 adult.

Injecting Large Constructs

The difficulty of obtaining transformants increases exponentially with the size of the DNA construct to be inserted. Constructs less that 20 kb in size may be injected relatively routinely, but increasing numbers of embryos need to be injected (and G1 flies screened) to obtain transformants carrying larger constructs. To obtain transformants of a 50-kb cosmid clone containing the entire Drosophila myosin heavy-chain gene, Cripps et al. (9) incorporated the following modifications: *(i)* DNA concentrations were increased to 1 mg/mL for the transforming plasmid and 100 µg/mL for the helper plasmid; and *(ii)* needles of slightly larger bore than normal were used to allow the viscous DNA solution to flow freely from the needle (15). In excess of 6000 embryos were injected, and three stable transgenic lines were isolated.

SOLUTIONS AND REAGENTS

DNA Injection Buffer

DNA for injection should be dissolved in a solution containing 5 mM KCl and 100 µM sodium phosphate buffer, pH 6.8.

Embryo Collection Solution

Embryos are washed from the grape/agar plate with a solution of 0.7% (wt/vol) NaCl and 0.03% (vol/vol) Triton X-100.

G418 Solution and Food

To select for G1 individuals carrying *neo*[R] as the selectable marker, G0 flies are crossed and placed in vials of food made with water containing the drug G418. Only transformed individuals will survive this drug, however differences in levels of expression between different lines might result in some lines being more resistant than others. G418 can be obtained from Life Technologies (Rockville, MD, USA) and is called Geneticin. It is shipped as a solid with an activity of 500 to 600 μg/1 mg. The working concentration is 500 μg/mL in water; therefore, a 1 mg/mL stock of the solid will give approximately the correct concentration of active drug. Food containing G418 can be made using Instant Fly Food from Carolina Biological Supply Company (Burlington, NC, USA). Simply place a scoop of the dried food mixture into a vial and add sufficient G418 solution to hydrate the food. Let stand for 5 minutes before adding flies. It is important that grains of yeast not be added, since this can interfere with the selection for resistant individuals.

ACKNOWLEDGMENTS

We are very grateful to Dr. Steve Stricker for invaluable assistance in preparing some of the figures. Research in the Cripps Laboratory is funded by grants from the Muscular Dystrophy Association and the American Heart Association, Desert/Mountain Affiliate. Research in the Bernstein Laboratory is funded by grants from the National Institutes of Health, the National Science Foundation, and the Muscular Dystrophy Association.

REFERENCES

1.Adams, M.D., R.S. Tarng, and D.C. Rio. 1997. The alternative splicing factor PSI regulates P-element third intron splicing in vivo. Genes Dev. *11*:129-138.
2.Anderson, K.V. and C. Nusslein-Volhard. 1984. Information for the dorsal—ventral pattern of the *Drosophila* embryo is stored as maternal mRNA. Nature *311*:223-227.
3.Ashburner, M. 1989. Drosophila, a Laboratory Handbook. CSH Laboratory Press, Cold Spring Harbor, NY.
4.Beall, E.L. and D.C. Rio. 1997. *Drosophila* P-element transposase is a novel site-specific endonuclease. Genes Dev. *11*:2137-2151.
5.Beall, E.L. and D.C. Rio. 1998. Transposase makes critical contacts with, and is stimulated by, single-stranded DNA at the P element termini in vitro. EMBO J. *17*:2122-2136.
6.Bellen, H.J., C.J. O'Kane, C. Wilson, U. Grossniklaus, R.K. Pearson, and W.J. Gehring. 1989. P-element-mediated enhancer detection: a versatile method to study development in *Drosophila*. Genes Dev. *3*:1288-1300.
7.Brand, A.H. and N. Perrimon. 1993. Targeted gene expression as a means of altering cell fates and generating dominant phenotypes. Development *118*:401-415.
8.Cooley, L., D. Thompson, and A.C. Spradling. 1990. Constructing deletions with defined endpoints in *Drosophila*. Proc. Natl. Acad. Sci. USA *87*:3170-3173.
9.Cripps, R.M., K.D. Becker, M. Mardahl, W.A. Kronert, D. Hodges, and S.I. Bernstein. 1994. Transformation of *Drosophila melanogaster* with the wild-type myosin heavy-chain gene: rescue of mutant phenotypes and analysis of defects caused by overexpression. J. Cell Biol. *126*:689-699.
10.Driever, W., J. Ma, C. Nusslein-Volhard, and M. Ptashne. 1989. Rescue of bicoid mutant *Drosophila* embryos by bicoid fusion proteins containing heterologous activating sequences. Nature *342*:149-154.
11.Elefant, F. and K.B. Palter. 1999. Tissue-specific expression of dominant negative mutant *Drosophila* HSC70 causes developmental defects and lethality. Mol. Biol. Cell *10*:2101-2117.
12.Garza, D., M. Medhora, A. Koga, and D.L. Hartl. 1991. Introduction of the transposable element mariner into the germline of *Drosophila melanogaster*. Genetics *128*:303-310.

13. Golic, K.G. 1994. Local transposition of P elements in *Drosophila melanogaster* and recombination between duplicated elements using a site-specific recombinase. Genetics *137*:551-563.

14. Golic, K.G. and S. Lindquist. 1989. The FLP recombinase of yeast catalyzes site-specific recombination in the *Drosophila* genome. Cell *59*:499-509.

15. Haenlin, M., H. Steller, V. Pirrotta, and E. Mohier. 1985. A 43 kilobase cosmid P transposon rescues the *fs(1)K10* morphogenetic locus and three adjacent *Drosophila* developmental mutants. Cell *40*:827-837.

16. Hunter, C. P. 1999. Genetics: a touch of elegance with RNAi. Curr. Biol. *9*:R440-R442.

17. Kalb, J.M., A.J. DiBenedetto, and M.F. Wolfner. 1993. Probing the function of *Drosophila melanogaster* accessory glands by directed cell ablation. Proc. Natl. Acad. Sci. USA *90*:8093-8097.

18. Kamdar, P., G. Von Allmen, and V. Finnerty. 1992. Transient expression of DNA in *Drosophila* via electroporation. Nucleic Acids Res. *20*:3526.

19. Karess, R.E. and G.M. Rubin. 1984. Analysis of P transposable element functions in *Drosophila*. Cell *38*:135-146.

20. Kennerdell, J.R. and R.W. Carthew. 1998. Use of dsRNA-mediated genetic interference to demonstrate that frizzled and frizzled 2 act in the wingless pathway. Cell *95*:1017-1026.

21. Kidwell, M.G. 1977. Reciprocal differences in female recombination associated with hybrid dysgenesis in *Drosophila melanogaster*. Genet. Res. *30*:77-88.

22. Kidwell, M.G. and J.F. Kidwell. 1976. Selection for male recombination in *Drosophila melanogaster*. Genetics *84*:333-351.

23. Kunes, S. and H. Steller. 1991. Ablation of *Drosophila* photoreceptor cells by conditional expression of a toxin gene. Genes Dev. *5*:970-983.

24. Laski, F.A., D.C. Rio, and G.M. Rubin. 1986. Tissue specificity of *Drosophila* P element transposition is regulated at the level of mRNA splicing. Cell *44*:7-19.

25. Lee, C.C., E.L. Beall, and D.C. Rio. 1998. DNA binding by the KP repressor protein inhibits P-element transposase activity in vitro. EMBO J. *17*:4166-4174.

26. Lee, E.C., S.Y. Yu, X. Hu, M. Mlodzik, and N.E. Baker. 1998. Functional analysis of the fibrinogen-related scabrous gene from *Drosophila melanogaster* identifies potential effector and stimulatory protein domains. Genetics *150*:663-673.

27. Lieberfarb, M.E., T. Chu, C. Wreden, W. Theurkauf, J. P. Gergen, and S. Strickland. 1996. Mutations that perturb poly(A)-dependent maternal mRNA activation block the initiation of development. Development *122*:579-588.

28. Lin, D.M., V.J. Auld, and C.S. Goodman. 1995. Targeted neuronal cell ablation in the *Drosophila* embryo: pathfinding by follower growth cones in the absence of pioneers. Neuron *14*:707-715.

29. Lindsley, D.L. and G. Zimm. 1992. The Genome of *Drosophila melanogaster*. Academic Press, San Diego.

30. Martin, P., A. Martin, A. Osmani, and W. Sofer. 1986. A transient expression assay for tissue-specific gene expression of alcohol dehydrogenase in *Drosophila*. Dev. Biol. *117*:574-580.

31. Misquitta, L. and B.M. Paterson. 1999. Targeted disruption of gene function in *Drosophila* by RNA interference (RNA-i): a role for nautilus in embryonic somatic muscle formation. Proc. Natl. Acad. Sci. USA *96*:1451-1456.

32. Misra, S. and D.C. Rio. 1990. Cytotype control of *Drosophila* P element transposition: the 66 kd protein is a repressor of transposase activity. Cell *62*:269-284.

33. Moffat, K.G., J.H. Gould, H.K. Smith, and C.J. O'Kane. 1992. Inducible cell ablation in *Drosophila* by cold-sensitive ricin A chain. Development *114*:681-687.

34. O'Hare, K. and G.M. Rubin. 1983. Structures of P transposable elements and their sites of insertion and excision in the *Drosophila melanogaster* genome. Cell *34*:25-35.

35. O'Kane, C.J. and W.J. Gehring. 1987. Detection in situ of genomic regulatory elements in *Drosophila*. Proc. Natl. Acad. Sci. USA *84*:9123-9127.

36. Picard, G., J.C. Bregliano, A. Bucheton, J.M. Lavige, A. Pelisson, and M.G. Kidwell. 1978. Nonmendelian female sterility and hybrid dysgenesis in *Drosophila melanogaster*. Genet. Res. *32*:275-287.

37. Robertson, H.M. and W.R. Engels. 1989. Modified P elements that mimic the P cytotype in *Drosophila melanogaster*. Genetics *123*:815-824.

38. Robertson, H.M., C.R. Preston, R.W. Phillis, D.M. Johnson-Schlitz, W.K. Benz, and W.R. Engels. 1988. A stable genomic source of P element transposase in *Drosophila melanogaster*. Genetics *118*:461-470.

39. Roche, S.E., M. Schiff, and D.C. Rio. 1995. P-element repressor autoregulation involves germ-line transcriptional repression and reduction of third intron splicing. Genes Dev. *9*:1278-1288.

40. Rosenberg, U.B., A. Preiss, E. Seifert, H. Jackle, and D.C. Knipple. 1985. Production of phenocopies by Kruppel antisense RNA injection into *Drosophila* embryos. Nature *313*:703-706.

41. Rubin, G.M., M.G. Kidwell, and P.M. Bingham. 1982. The molecular basis of P-M hybrid dysgenesis: the nature of induced mutations. Cell *29*:987-994.

42. Schuh, R. and H. Jackle. 1989. Probing *Drosophila* gene function by antisense RNA. Genome *31*:422-425.

43. Sharp, P.A. 1999. RNAi and double-strand RNA. Genes Dev. *13*:139-141.

44. Spradling, A.C. and G.M. Rubin. 1982. Transposition of cloned P elements into *Drosophila* germ line chromosomes. Science *218*:341-347.

45. Sved, J.A. 1976. Hybrid dysgenesis in *Drosophila melanogaster*: a possible explanation in terms of spatial organization of chromosomes. Aust. J. Biol. Sci. *29*:375-388.

46. Tabara, H., A. Grishok, and C.C. Mello. 1998. RNAi in *C. elegans*: soaking in the genome sequence. Science *282*:430-431.

47. Timmons, L., J. Becker, P. Barthmaier, C. Fyrberg, A. Shearn, and E. Fyrberg. 1997. Green fluorescent protein/beta-galactosidase double reporters for visualizing *Drosophila* gene expression patterns. Dev. Genet. *20*:338-347.

48. Tower, J., G.H. Karpen, N. Craig, and A.C. Spradling. 1993. Preferential transposition of *Drosophila* P elements to nearby chromosomal sites. Genetics *133*:347-359.

49. Vanario-Alonso, C.E., E. O'Hara, W. McGinnis, and L. Pick. 1995. Targeted ribozymes reveal a conserved function of the *Drosophila* paired gene in sensory organ development. Mech. Dev. *53*:323-328.

50. Wilson, C., R.K. Pearson, H.J. Bellen, C.J. O'Kane, U. Grossniklaus, and W.J. Gehring. 1989. P-element-mediated enhancer detection: an efficient method for isolating and characterizing developmentally regulated genes in *Drosophila*. Genes Dev. *3*:1301-1313.

51. Zhao, J.J. and L. Pick. 1993. Generating loss-of-function phenotypes of the *fushi tarazu* gene with a targeted ribozyme in *Drosophila*. Nature *365*:448-451.

52. Zuker, C.S., D. Mismer, R. Hardy, and G.M. Rubin. 1988. Ectopic expression of a minor *Drosophila* opsin in the major photoreceptor cell class: distinguishing the role of primary receptor and cellular context. Cell *53*:475-482.

7 Introduction of DNA into Mouse Embryos

Elizabeth L. George and David S. Milstone
*Vascular Research Division, Department of Pathology,
Brigham and Women's Hospital and Harvard Medical School,
Boston, MA, USA*

INTRODUCTION

The ability to introduce cloned DNA into mouse embryos has resulted in countless novel and informative mutant strains of mice. These animals have led to discoveries in all areas of cell biology, physiology, and development, which in turn have greatly contributed to our understanding of disease. We will present several techniques, and variations of techniques, used to introduce cloned DNA into embryos and point out how the practical characteristics of each determine their individual utility. We hope this will convey a clear view of how this broad experimental approach is currently applied to a wide range of scientific investigation and how additional technical advances should lead to even more far-reaching applications in the near future.

Initially, cloned DNA was introduced by microinjection into one-cell embryos, leading to nondirected integration into the genome. These genome insertions are usually referred to as exogenous transgenes. The next two major advances relate to uses of murine embryonic stem (ES) cells. These cells retain totipotency in culture and can therefore act as a bridge between in vitro manipulation of the genome and in vivo analysis using whole animals. Also, ES cells undergo genetic recombination between transfected DNA and homologous genomic loci at a relatively high frequency. This ability to transfect cloned DNA into ES cells, and to detect rare homologous recombination events in individual clones in culture, is the basis for what are now referred to as endogenous gene modifications (or gene targeting). Most recently, exogenous transgenes and endogenous gene modification have been combined to allow conditional mutagenesis. This powerful combination has enabled generation

Gene Transfer Methods
Edited by Pamela A. Norton and Laura F. Steel
©2000 Eaton Publishing, Natick, MA

127

of both cell-type-specific mutations and subtle intragene mutations.

In Table 1, we outline the general steps and the minimum time required to generate novel mouse strains by the two general approaches: *(i)* introducing an exogenous transgene, and *(ii)* modifying an endogenous gene. The stated time requirements are for background knowledge and budgeting issues only and are not intended to serve as a basis for choice of experimental approach. While both approaches have a wide range of applications, each also has its own methodologic and scientific limitations, as we hope to convey in the following more detailed discussion.

Designing exogenous transgene vectors and the equipment and procedures used to introduce DNA and ES cells into embryos have largely become standardized, and several excellent reviews of these topics have been published (20,24,26,43,61). We will not repeat this information, but will focus on comparing exogenous transgenes and endogenous gene modifications.

INTRODUCING EXOGENOUS TRANSGENES

Introduction of exogenous DNA into embryos leads to nontargeted integration into chromosomal DNA(43). As a result, each transgenic line is different from every other (see below), and all phenotypes must be verified in mice derived from at least two independent founder animals. The experimental goal is usually over-expression of the transgene, which may encode a gene product that is not normally present in the target tissue (20). Expression of candidate oncogenes to deregulate cell growth control and produce tumors in transgenic animals was a popular early example of this application. A transgene may also be expressed at abnormally elevated levels, or in cell types, developmental stages, or physiologic stages in which the corresponding endogenous gene is normally silent. This provides a powerful approach to investigate altered expression of normal gene products in vivo. Other early studies focused on obtaining transgene expression that faithfully mimicked the expression patterns of endogenous genes. This involves defining the genetic regulatory sequences required for expression and including these in the DNA transgene introduced. Conditional expression requiring mating between activator and responder strains of transgenic mice is complex and not widely used at the present time (15,42). Innovative, pharmacologically based conditional transgene expression systems are promising but have also not been widely implemented (56).

Advantages of Exogenous Transgenes

If copy number-dependent and position-independent expression is achieved (see below), then several practical advantages of over-expression by microinjection of DNA are immediately apparent. Transgenic animals can be

Table 1. Minimum Time Required to Produce Mice Bearing Exogenous Transgenes or Targeted Mutations

Introducing Exogenous Transgenes	
Week	**Procedure**
variable	Assemble gene expression cassette.
variable	Test expression in vitro.
1	Microinject one-cell embryos. Oviductal adoptive transfer to pseudo-pregnant recipients.
3	Founder generation born.
9	Founder mice mated to inbred partners at 6 weeks of age to establish independent transgenic lines.
12	First generation of transgenic lines born.
18	First generation of transgenic lines reach adult maturity at 6 weeks of age.

Modifying Endogenous Genes	
Week	**Procedure**
variable	Assemble gene targeting vector.
1 to 6	Transfect ES cells.
	Identify, isolate, and preserve targeted clones.
7	Blastocyst injection or morula aggregation of ES cells. Uterine adoptive transfer to pseudo-pregnant females.
9	Chimeric generation born.
15	Chimeric generation mated to inbred partners at 6 weeks of age.
17	Heterozygous generation born. Identify those with the mutation transmitted through the germline.
24	Heterozygotes intercrossed at 6 weeks of age.
27	First generation containing homozygotes born.
34	Homozygotes reach adult maturity at 6 weeks of age.

produced much more rapidly by this approach than by homologous recombination in ES cells (Table 1). This is important especially if many genetic modifications will be evaluated. In addition, transgene copy number frequently varies more than 100-fold between different founder animals derived by microinjection. This, and position effects, results in transgenic lines with a wider range of expression levels than is possible using targeted mutagenesis. Dose-response relationships are thus more easily tested over a wider range in vivo.

Obtaining a desired transgene expression pattern is predicated on the availability of appropriate genetic regulatory elements. Vector sequences often interfere with expression and should be removed from the transgene before microinjection. Including an intron with appropriate splice acceptor and donor sites often enhances transgene expression but, in other situations, can be detrimental (9,44). The effect of incorporating foreign DNA in an exogenous transgene is, therefore, unpredictable. This is especially important for transgenes incorporating cDNAs. Several generic transgene constructs have been designed for expressing various cDNA coding sequences under the control of various regulatory elements. Examples include: *(i)* human growth hormone gene fragments containing exons and introns that result in RNA splicing combined with a mutation that leads to biologically inactive growth hormone polypeptide (44); and *(ii)* a synthetic hybrid intron combining an adenovirus splice donor and an immunoglobulin splice acceptor site (9).

Exogenous transgenes can test the activities of gene regulatory elements in vivo. The usual goal is copy number-dependent, position-independent expression in the appropriate cell types and at the appropriate times during development. This pattern implies that the minimal regulatory elements necessary for authentic expression are contained in the injected DNA. Because transgenes as large as several hundred kilobase pairs, or even entire bacterial artificial chromosomes (BACs) or yeast artificial chromosomes (YACs), can be injected routinely, it is often possible to obtain authentic expression even when the required regulatory elements have not been localized. Deletion and site-directed mutagenesis can then be used to define the regulatory elements present in the transgene. The activity of single regulatory elements, as well as combinations and altered spatial arrangements of multiple elements, can also be determined. Such analyses are usually more readily accomplished using exogenous transgenes than endogenous gene modifications.

Methods of Introducing Exogenous Transgenes into Embryos

Microinjection of one-cell embryos is the approach most commonly used for introducing exogenous transgenes. Toxicity of DNA to one-cell embryos is a crucial issue and may require screening different preparations of the DNA to be injected. Specific methods for selecting embryo donor mice, preparing and injecting embryos, and reimplanting embryos into foster mothers have been well described previously (24,48).

Retroviral-mediated gene transfer was popularized earlier than microinjection, but has more limited utility. Retroviruses integrate into chromosomal DNA with very high efficiency and are therefore a method of choice for cell lineage analysis (8). However, transgenes introduced using retroviruses are often not expressed in predictable or consistent patterns.

Adenovirus-mediated gene transfer into mid- and late-stage murine embryos has also been successful and can lead to experimentally useful transgene expression (4,31). However, integration into chromosomal DNA generally does not occur, expression is transient, and heritable genetic modifications are not achieved. In addition, concerns regarding the eventual development of immune reactivity to adenoviral components have not been fully resolved, although neonatal tolerance may render mice derived from infected embryos relatively resistant to this complication.

Exogenous transgenes may also be introduced into ES cells by Ca_2Cl-mediated DNA precipitation and lipid-based methods. These approaches have been used to express genes normally silent in ES cells in vitro and to introduce rescue constructs for in vitro analysis of hematopoiesis (55,65), but have not been widely used for germline transmission.

Mechanisms of Exogenous Transgene Integration into Genomic DNA

Chromosomal integration of microinjected DNA usually occurs soon after injection, resulting in a stable transgenic allele in all cells of the founder animal. However, approximately 10% to 20% of founder mice are chimeras of transgenic and nontransgenic cells, suggesting that integration was delayed until at least the two-cell stage of embryogenesis. The transgenic locus usually consists of head-to-tail concatamers of as many as several hundred copies of the injected DNA (5). Deletion, replication, and/or inversion of genomic DNA commonly occur adjacent to the integration site, and approximately 10% to 30% of transgenic alleles cause recessive or dominant mutations. This emphasizes that microinjection of DNA is mutagenic, which must be considered when interpreting results. Every phenotype must be confirmed in mice derived from two or more independent founder animals. Otherwise, one cannot rule out the possibility that a phenotype reflects insertional mutation rather than presence of the transgene.

MODIFYING ENDOGENOUS GENES

Prior to 1987, when Thomas and Capecchi (59) first inactivated a gene in the mouse embryonic stem cell genome by homologous recombination, endogenous gene modifications in mammals were extremely rare. Only spontaneous mutations that gave rise to obvious phenotypes, such as those comparable to heritable human diseases, were available for study. The ability to make directed mutations in the mouse genome, by performing homologous recombination in ES cells, changed the situation dramatically. This strategy has become fairly common (37), and is now referred to as gene targeting. If cultured properly, ES cells retain their totipotency and can contribute to the germ cell lineage even after the extensive manipulations required to achieve homologous recombination. Inactivation of a gene by homologous recombination, followed by generation of a mutant mouse strain in which all cells of the animal harbor the mutation, is now referred to as a classical gene knockout.

In many cases, a null mutation results in embryonic lethality. This identifies the first stage in embryogenesis at which that gene is essential and limits the experimental analysis to developmental processes. However, a gene may also be involved in earlier processes, but may not be essential (10,25). A classical gene knockout of the extracellular matrix component fibronectin, for example, results in lethality during gastrulation (16). This precludes evaluating physiologic and disease models, such as fibronectin's proposed role in initiation of fibrosis following liver injury (27). In contrast, a gene knockout of thrombospondin-1 results in viable mice, even though expression patterns predict important roles for this gene product during development (30). Another surprisingly mild developmental phenotype is seen in the alpha V integrin knockout (3). Many of these mutant mice survive until birth, with only limited vascular defects. Given the in vitro data concerning this integrin and angiogenesis (6), more severe defects had been predicted. Finally, there are also many examples of genes for which knockouts were not predicted to be lethal, such as the selectin gene family. Knockout mice are viable and fertile (34,36), even when two or all three of the family members are inactivated (14,49).

General Design of Gene Knockout Vectors

The general strategy for design of a gene knockout targeting vector is diagrammed in Figure 1. As a first consideration, genomic clones should be isogenic with the line of ES cells to be used, most of which are derived from strain 129/Sv mice. This, together with longer lengths of homologous sequence (generally 3–4 kb) in the targeting vector, has been shown to increase the frequency of homologous recombination (12). The approach most likely to succeed is to delete the translation initiation signal and some mature protein coding sequence and replace these with a positive selection gene. The two most commonly used positive selection genes confer resistance to G418 or hygromycin B by replacing the bacterial control sequences of the *neo* or *hyg* genes with mammalian control sequences. The most commonly used promoter and polyadenylation signal are from the mouse phosphoglycerate kinase gene (PGK) (1).

To enrich for homologous recombination events, a negative selection gene can be added to either end of the targeting vector. The most common gene used in ES cells, the herpes simplex virus thymidine kinase gene (52), renders cells sensitive to gancyclovir or FIAU (1-[2′-deoxy-2′-fluro-β-D-arabinofuranosyl]-5-iodouracil). In many cases, while the use of a negative selection gene does improve the recovery of homologous recombinants, the magnitude of this increase is not sufficient to justify the additional labor involved in construction of the targeting vector.

Initial screening of transfected ES cells for the targeted allele requires care-

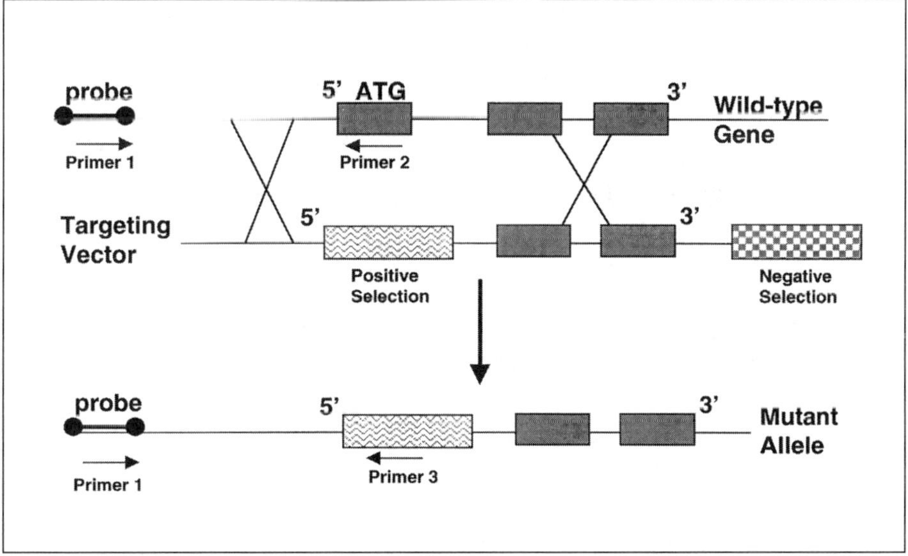

Figure 1. **Vector design for a classical gene knockout.** Deletion of a segment of a wild-type gene and replacement with a positive selection gene (wavy lines) is essential for generation of a gene targeting vector. Inclusion of a negative selection gene (checkered) at either end is optional, but may prove useful in reducing the number of colonies that must be screened. A suitable genomic probe, which lies outside the targeting vector, is essential for screening by genomic Southern blot. Such a probe is still necessary for subsequent characterization of targeted clones, even if screening is performed by PCR. For initial screening by PCR, three primers must be characterized, which will allow amplification of both the wild-type and targeted alleles. Primer 1 lies out side of the targeting construct, and is used to amplify both alleles. Primer 2 is specific to the region of the gene to be deleted. Primer 3 is specific to the positive selection gene. Primers must be selected to produce amplification products that are distinct in size and small enough to be amplified efficiently from relatively crude preparations of ES cell DNA. Horizontal lines represent intron or flanking sequences. Boxes represent exons (solid shading) or selection genes (patterned shading). Horizontal arrows represent PCR primers and the direction of extension. Crossed line represent possible sequences involved in homologous recombination. One of the two recommended Southern blot probes is indicated 5′ to the targeting construct.

ful and thorough characterization of the hybridization probe and/or polymerase chain reaction (PCR) primers to be used. This should be completed and verified while designing and constructing the targeting vector and before transfection. Modifications may have to be made to achieve adequate probe and/or amplimer length without sacrificing specificity. A probe for genomic Southern blot analysis must lie outside the targeting construct and must be tested to be sure it does not hybridize with repetitive sequences. PCR primers must span the junction between the targeting construct and genomic DNA. Lastly, the targeting vector must include a unique restriction site for linearization before electroporation. Many vectors are commercially available that include 8-bp recognition sequences for restriction enzymes such as *Not*I.

Homologous recombination in ES cells and blastocyst-mediated transgenesis can also be used to express exogenously introduced DNA via the "knockin" approach. This involves inserting exogenous DNA into an exon of a gene of interest, such that the exogenous DNA will also be transcribed whenever the targeted endogenous gene is expressed. Expression of the endogenous gene product may remain unaffected, especially if an internal ribosome entry site is incorporated into the targeted allele, and if markers used to select transfected ES cell clones are removed using, for example, Cre recombinase. Faithful expression usually does not require that genetic regulatory elements be identified beforehand, but only that the targeted allele not interfere with the endogenous expression pattern. This can be a great advantage if the experimental goal is limited to mimicking the expression of an endogenous gene and/or if the regulatory elements required for faithful expression have not been identified (which is usually the case).

Pluripotent Embryonic Stem Cells

Details of the various protocols used to derive, maintain, and use ES cells seem less important than the history of the particular cell line that one is working with. Several variations of common protocols have been successful, and stating that any given procedure must be performed in a specific way to ensure success seems unwarranted. Instead, it is more important to follow the particular protocol that has been used successfully with the specific cell line that one is working with. One must also be familiar with the growth characteristics of the cell line being used. Parameters such as growth rate, passage interval, optimal overall culture density at time of passage, and tolerance of suboptimal growth conditions vary greatly between lines. The size and morphology of ES cell colonies at the time of passage are usually critical parameters that predict the health, degree of differentiation, and probability of germline transmission of a given ES cell line. Excellent examples and discussions of optimal colony criteria have been published elsewhere (24,48), and additional information with illustrations is available via the Internet. One example of the latter can be found at (**http://www.medicine.wustl.edu/~es-core/**), and many more can be easily located using Internet search engines.

ES cells have been derived using many variations of a common protocol (24,48). In general, embryos are removed at the delayed blastocyst stage from ovariectomized pregnant females and cultured on mitotically inactivated feeder (murine embryo feeder; MEF) cells prepared from E13.5-E14.5 embryos. LIF (leukemia inhibitory factor, also known as DIA, differentiation inhibitory activity) is produced by MEF and helps maintain the pluripotent state of ES cells. Additional LIF is also frequently supplemented from purified sources or as conditioned medium from LIF-expressing cell lines. When the inner cell mass (ICM) begins to proliferate, the colony is passaged and expanded. The established ES cell line is cryopreserved by standard techniques, and aliquots are thawed to perform each subsequent experiment. It is not known whether such lines arise by clonal proliferation of a single ICM cell, or whether simultaneous proliferation of multiple initial cells is involved. ES cell lines can routinely be expanded several (at least 12) passages prior to cryopreservation and still maintain pluripotency after additional genetic manipulations and selection with cytotoxic drugs. In fact, pluripotent cell lines can frequently be recovered by cloning from an ES cell line that has undergone some degree of differentiation and has lost pluripotency.

One generic cell culture issue bears emphasis. It is essential to avoid mycoplasma contamination of ES cell cultures. Mycoplasma interferes with germline transmission and may modify in vitro differentiation. For these reasons, antibiotics and antimycotics should not be used for the routine passage of ES cells, especially early after isolation from the ICM. In addition, when possible, one should avoid sterilizing cell culture components by filtration. Antibiotics, antimycotics, and 0.2-μm filtration each suppress inadvertent bacterial and/or fungal contamination but do not eliminate mycoplasma. Drugs used for selection after transfection are unavoidable but can have similar consequences. Their application should be limited to cells destined for experimental use. When filtration is necessary, 0.1-μm filters should be used. It is, of course, preferable to avoid unnecessary exposure to potential sources of mycoplasma.

PROTOCOLS

We cannot overestimate the importance of adhering to the precise maintenance protocol developed for the particular ES cell line being used. This is probably the single most important factor determining the success or failure of germline transmission. Different lines show different tolerance of, for example, growth at relatively low or high cell density. However, some general principles apply to successful culture of virtually all ES cell lines. ES cells grow rapidly with a cell doubling time as short as 12 hours. Cultures must, therefore, be fed daily without fail to avoid unintended differentiation. The only exception should be when expanding cells for preparing DNA, when moderate differentiation is not detrimental. Because complete dissociation to single cells is essential, careful rinsing with divalent cation-free saline solutions prior to enzymatic treatment, and vigorous trituration afterward, are important. Many investigators report leaving ES cells in enzymatic digestion solutions for extended periods (up to 30 min) without apparent ill effects. While this is not recommended routinely, it does illustrate that ES cells survive, and may thrive, using cell dissociation conditions lethal for many other cell types. Some investigators remove residual traces of trypsin by centrifugation prior to plating ES cells in fresh medium, while others do not. MEF are usually plated the day prior to ES cell passage, but both cell types can be plated simultaneously with good results.

Growth and Maintenance of ES Cells

Some standard reagents used throughout this chapter are listed below. Appropriate cell culture medium and supplements can be obtained from several suppliers, such as Life Technologies (Rockville, MD, USA). Medium with high glucose concentrations are generally favored, N-[2-hydroxyethyl] piperazine-N'-[2-ethanesulfonic acid (HEPES) buffer is included to protect ES cells during prolonged manipulations in room air, and sodium pyruvate is sometimes added. LIF can be supplied as recombinant protein (ESGRO from Life Technologies) or as culture supernatant from CHO cells transfected with an expression construct for LIF (see below). Note that G418 is supplied as a mixture of two stereoisomers, only one of which is biologically active. The active concentration is thus one half of the total G418 concentration.

The quality of fetal bovine serum (FBS) used is critical, and individual serum lots should be screened on the specific ES cell line to be used in experiments. The ultimate criteria are germline transmission and genomic integrity (nonmutagenic), but evaluation of these parameters is time-consuming. ES cell plating efficiency, colony morphology, growth rate, and cellular differentiation are usually substituted as criteria for selecting serum. If all other parameters are equal, then lower levels of endotoxin are preferable. Hyclone Laboratories (Logan, UT, USA) has been one consistently reliable source. The time and expense of screening serum can be minimized by using nonscreened, but otherwise high quality, lots of heat-inactivated (HI)-FBS to prepare MEF. Screened lots can then be reserved for use only when ES cells are present.

<u>Complete Medium:</u>

Dulbecco's modified Eagle medium (DMEM)	—
HI-FBS	15% (vol/vol)
HEPES, pH 7.4	25 mM
Glucose	4.5 g/L
L-glutamine	2 mM
β-mercaptoethanol	50 µM
Nonessential amino acids	0.1 mM each amino acid
Sodium pyruvate	1 mM

Antibiotics or antimycotics should not be included for routine maintenance of ES cells. All medium should be made using a single lot of pre-screened HI-FBS for successful ES cell maintenance. ES cells are cultured at 37°C in a humidified atmosphere of 5% CO_2 in air. ES cell culture on feeder cell layers has traditionally been performed using tissue culture plasticware coated with gelatin. Cover the growth surface of culture vessels with 0.1% gelatin and incubate for 20 to 60 minutes at room temperature or 37°C. These vessels may be stored at 4°C, or in a 37°C humidified incubator, until needed. Alternatively, the gelatin may be aspirated, and the vessels may be stored dry until used. Some investigators have successfully omitted gelatin coating altogether.

<u>LIF:</u> This can be obtained as a purified protein preparation produced by recombinant techniques (ESGRO; 10 000 U/mL stock concentration) and is included in the culture medium at 500 to 1000 U/mL to prevent differentiation of ES cells. One unit of LIF activity is defined as the amount, when present in a 1 µL of medium, sufficient to prevent ES cell differentiation. Alternatively, one can use between 1:250 and 1:1000 dilutions of a culture supernatant from CHO cells bearing a LIF expression construct (Genetics Institute, Cambridge, MA, USA).

<u>MEF cells:</u> These can be prepared, by well-described techniques (24), from E13.5 to E14.5 embryos of several lines of mice bearing targeted mutations incorporating the resistance markers, such as *neo* or *hyg*, to be used in specific experiments. A ubiquitous transcriptional promoter usually drives resistance marker expression, but the actual specificity of expression is usually not known and may vary between mutant mouse lines. Appropriate lines are generally available for this purpose from numerous laboratories or can be purchased. A line of mice bearing several resistance markers has also been described (60). Gamma irradiation (3000 RAD) or mitomycin C treatment is used to prevent MEF proliferation. Convenient single-use aliquots of mitotically inactive MEF can be cryopreserved in advance and thawed as needed.

MEF are plated at sufficient densities to form a confluent, but not over-crowded, monolayer. Experience will dictate the appropriate density. MEF are often plated one day prior to adding ES cells but both can also be plated simultaneously. MEF can also be used up to 3 or 4 days after plating, although the ES cells cultured on them should be passaged again relatively soon. Medium made with nonscreened lots of HI-FBS, and without LIF, can be used to maintain MEF until ES cells are added.

<u>HEPES-Buffered Saline (HBS)</u>: ES cells are rinsed prior to enzymatic dissociation in a buffered mixture of physiologic salts and EDTA prepared from standard reagents.

Combine:

> 8.8 mL of a 10× balanced salt solution (such as Hank's Balanced Salts)
> 2 mL 1 M HEPES (pH 7.3)
> 170 µL 1 N NaOH and 70 mL water
> Adjust the final pH to 7.4 with NaOH or HCl. PBS without Ca^{++} or Mg^{++} can be substituted for HBS.

<u>G418 Medium and Stock Solutions</u>: G418 is acidic and is supplied as an equimolar/equimass mixture of active and inactive isomers. This does not create problems when used at the standard concentration of 0.2 mg active G418/mL to select ES cells expressing the *neo*^{R} selectable marker. Convenient aliquots of G418 (e.g., 200 mg total, 100 mg active, for 500 mL medium) can be weighed beforehand, stored at room temperature protected from light, and dissolved in the appropriate volume of medium prior to use. However, when using higher concentrations of G418 to select for loss of heterozygosity (LOH) (see below) some special handling is required.

> Weigh G418 (Life Technologies) into a sterile 50-mL polypropylene tube. Add DMEM (without HEPES) at 1 mL/800 mg total G418 and mix well to dissolve.
> Neutralize carefully to a pH of approximately 7.0 using 10 N NaOH added in small aliquots (<20 µL). This is essential to avoid pH changes during selection.
> Add DMEM (without HEPES) to a final concentration of 200 mg active (400 mg total) G418/mL.
> Sterilize by 0.1-µm filtration and store 0.5-mL aliquots at -20°C.

<u>FIAU or Gancyclovir Stock Solutions</u>: The working concentrations of FIAU and gancyclovir are 0.2 and 2.0 µ M, respectively. Stock solutions are 10 mM in PBS. FIAU may not be available, but gancyclovir can be substituted with success.

Electroporation Buffer:

Combine:

49 mL Dulbecco's PBS, without Ca^{+2} or Mg^{+2}

1 mL 1 M HEPES (pH 7.3)

91 µl 55 mM β-mercaptoethanol (final concentration is 100 µM)

10 mM EDTA/HBS (*Versene solution may be substituted*):

Weigh $EDTA(Na)_2$-2 H_2O into a sterile 50-mL polypropylene tube. Add sufficient HBS to bring the $EDTA(Na)$-2 H_2O to 3.72 mg/mL (10 mM final concentration). Mix to solubilize.

Neutralize this solution carefully to pH approximately 7.4 using 1 *N* NaOH added in small aliquots. This will require approximately 1.1 mL 1 *N* NaOH per 100 mL 10 mM EDTA/HBS. Monitor the pH by eye using the phenol red indicator dye in the HBS. Confirm these observations using a pH meter.

Trypsin/EDTA/HBS:

Combine:

1 part stock solution of cell culture grade trypsin (2.5%)

8 parts HBS

1 part 10 mM EDTA/HBS

Alternatively, dilute trypsin stock to 0.25% in versene.

Proteinase K Digestion Buffer:

Combine:

10 mL of 5 M NaCl (final concentration 100 mM)

5 mL of 1 M Tris- HCl, pH 8.0 (final concentration 10 mM)

25 mL 0.5 M EDTA, pH 8.0 (final concentration 25 mM)

12.5 mL 20% sodium dodecyl sulfate (SDS) (final concentration 0.5%)

448 mL water

Add proteinase K to a final concentration of 0.1 mg/mL. After addition of proteinase K, this reagent can be stored for at least several weeks at -20°C. Avoid multiple freeze/thaw cycles of this reagent.

We will describe our modified version of protocols that have been successful in a number of laboratories using the J1 embryonic stem cell line derived by Drs. En Li, Rudolf Jaenisch, and coworkers (32). However, different established cell lines may perform less well using these protocols, which should thus be considered illustrative only. We will concentrate on the overall technical and experimental strategies that maximize the efficiency of germline modification in mice.

If one is not successful using an established ES cell line, even after scrupulously emulating the procedures used by others working with that line, then one should consider deriving a new ES cell line starting from the blastocyst

ICM. The experience already acquired by culturing the established ES cell line will improve the chances of deriving a new line that performs well under the conditions used.

1. Plate sufficient MEF to produce a confluent monolayer in medium without LIF. Approximately 3×10^6 MEF cells are usually sufficient for one 10-cm dish.

2. On the following day, feed the ES cells to be passaged with fresh medium/LIF 2 hours prior to dissociation.

3. Aspirate medium from ES cells and rinse twice with 5 to 10 mL HBS.

4. Add 2 mL trypsin/EDTA/HBS and incubate 3 minutes at 37°C in a CO_2 incubator.

5. Triturate 10 times with a micropipet tip on the end of a 5-mL pipet, or sterile cotton-plugged Pasteur pipet, and view the cells with a phase-contrast microscope. If they are not largely single cells, return them to the incubator for an additional 3 minutes, repeat the trituration, and view them again. This is very important, as complete dissociation prior to passage is essential for maintaining the undifferentiated phenotype.

6. Add 3 mL medium and triturate 10 times with a micropipet tip on a 5-mL pipet. Dilute the cells and count them using a hemacytometer. The yield should be approximately 5×10^6 to 2×10^7 cells from each 100- or 150-mm plate.

7. Plate 0.5 to 2×10^6 cells in 10 to 15 mL media in 100-mm plates of MEF or 1 to 4×10^6 cells in 20 to 30 mL media in 150-mm plates (see note below).

8. Feed cultures daily with the same volume of medium/LIF and passage as required.

<u>Note:</u> Cultures are often passaged based on visual estimation of colony size and density, rather than on actual cell counts, as detailed above. In this case, cultures are usually diluted 5- to 6-fold every 2 to 3 days. However, optimal passage frequencies and ratios vary significantly between different ES cell lines. It is therefore essential to know the preferred passage schedule and cell density of the particular ES cell line with which one is working. Sparse cultures should be passaged based on the appearance of individual colonies while avoiding overgrowth and differentiation (see the discussion of expanding drug-resistant clones below).

Deriving ES Cells from Delayed Blastocysts

Techniques for isolating ES cell lines from delayed blastocysts are outlined below. This approach can be successful even if one has encountered difficulty maintaining a line initially established in another laboratory. Ovariectomy and delayed implantation lead to significantly greater success. This is definitely worth the modest effort required, even for those unfamiliar with survival surgery in small rodents.

Day -1	Mate mice.
Day 0	Detect copulation plug.
Day 2	Perform ovariectomy and administer DepoProvera (Sigma, St. Louis, MO, USA) (see References 24 and 48).
Day 6	Flush blastocysts from uteri and culture each in an individual C24 well of MEF in 1 mL ES cell medium/LIF. Leave cultures undisturbed in an incubator for 2 days.
Day 8	Examine cultures to determine if embryos have attached. Carefully feed attached embryos with fresh medium/LIF.
Day 9	Feed attached embryos with fresh medium/LIF.
Day 10	Rinse adherent embryos twice with 1 mL HBS and leave them in 1 mL HBS. Pick each ICM using an autoclaved DNA sequencing gel loading pipet tip, or other thin pipets (drawn-out, autoclaved 50-μL glass pipet). Transfer each ICM to 50 μL of 0.25% trypsin, 1 mM EDTA in a round-bottom microplate well and incubate 3 minutes at 37°C in a 5% CO_2 incubator.
	Dissociate the trypsinized ICM to small aggregates (clumps of between 4 and 10 cells) by repetitive pipeting with a thin pipet tip and plate these on fresh MEF in C24 wells containing 1 mL medium/LIF. Do not dissociate the ICM to a uniform single cell suspension at this stage as this may decrease the yield of ES cell lines.
Day 11 to 15	Feed daily with 1 mL of medium/LIF and examine for growth of ES cell colonies, colonies with other morphologies, and mixed colonies (24,48).
Day 15	Rinse each well twice with 1 mL HBS. Add 1 drop trypsin/EDTA from a 2-mL pipet and incubate for 3 minutes at 37°C in 5% CO_2. Add 1 mL medium/LIF, triturate at least 10 times with a micropipet tip secured on the end of a 5-mL disposable sterile pipet, and examine the result with a phase-contrast microscope.
	If cells are a uniform single cell suspension, then plate them in individual gelatinized C12 wells containing 1 mL medium/LIF (without additional MEF). If significant cellular aggregates remain, then repeat trituration,

	re-examine, and plate when dissociation is complete.
Day 16 and on	Feed daily with 2 mL of medium/LIF. ES cell colonies are usually seen by Day 17.5. Passage these to individual C6 wells of fresh MEF in 3 mL medium/LIF and feed daily. Rinse twice with 3 mL HBS, add 0.5 mL trypsin/EDTA/HBS, and incubate for 3 minutes at 37°C in 5% CO_2. Add 1.5 mL medium and triturate to obtain a single cell suspension. Centrifuge at 200× *g* for 5 minutes at room temperature and resuspend the cell pellet in 1.25 mL cold medium. Slowly add 1.25 mL cold 20% dimethyl sulfoxide (DMSO) in medium and freeze 0.5-mL aliquots in LN2 as established ES cell lines. Such lines are usually 4 passages after initial outgrowth from the ICM. If one assumes a clonal origin, then the approximate number of cell doublings involved can be calculated, if desired.

Transfection and Screening ES Cells

Transfection of ES cells is generally performed by electroporation, because this favors integration of a single copy of the targeting construct into a single genomic locus. Delaying the introduction of drugs for positive and negative selection by 20 to 21 hours and 48 hours, respectively, is important to allow expression of the selectable marker proteins from the newly integrated resistance genes. One should use the minimal concentration of selection drugs that give efficient killing of nontransfected cells. This allows mutagenesis at numerous different loci, using a single drug concentration, even though the expression of resistance markers may vary between integration sites.

1. Prepare high-quality plasmid DNA of the targeting construct by any of several common techniques. Qiagen (Valencia, CA, USA) columns or two rounds of equilibrium density ultracentrifugation in CsCl gradients work well (see Chapter 2).

2. Linearize 100 to 200 µg of targeting construct plasmid DNA with an appropriate restriction enzyme.

3. Analyze 0.25 µg of the linearized construct on an agarose/ethidium bromide gel with size and quantitation standards to verify complete cutting.

4. Extract the linearized construct twice with phenol:CHCl3 (1:1) and twice with CHCl3.

Perform all subsequent steps, except centrifugations, in a laminar flow environment and use only sterile reagents specifically reserved for ES cell culture.

5. Precipitate the extracted construct in a microcentrifuge tube by routine methods using EtOH or isopropanol and adding NaCl, if necessary. Centrifuge 5 minutes at 15 000× g at room temperature.

6. Wash the pellet twice with an equal volume of 70% EtOH or 50% isopropanol. Remove traces of alcohol by leaving the uncapped tube open in the hood, but do not evaporate to dryness.

7. Resuspend the pellet in tissue culture grade water or electroporation buffer at a predicted concentration of 1 μg of linearized construct in 1 μL of solution.

8. Dilute 1 μL of the resuspended construct with 9 μL of water. Analyze 1 μl of this dilution on a gel with size and quantitation standards to ensure that the construct is not degraded. (Optional: dilute 2 μL in 1 mL of water and determine A260/A280). Estimate the DNA concentration and adjust to 1 μg/μL, if necessary.

9. Store the construct at 4°C for 1 to 2 weeks or at -80°C for longer periods. Avoid numerous freeze/thaw cycles.

10. On the day before electroporation, prepare five 100-mm plates of MEF in complete medium without LIF.

11. Dissociate the ES cells to be transfected exactly as described above for routine maintenance.

12. Centrifuge 1×10^7 cells at 100 to 200× g for 5 minutes at room temperature.

13. Wash the cell pellet twice in 2 mL of electroporation buffer, centrifuging as in Step 12 between washes.

14. Resuspend the cell pellet in 1 mL of electroporation buffer.

15. Transfer 0.8 mL of this washed cell suspension to one 0.4-cm electroporation cuvette. Add 20 μg of targeting construct and mix well using a 2-mL pipet.

16. Electroporate at room temperature following the instructions accompanying the electroporation device used. Common conditions using the Gene Pulser (Bio-Rad, Hercules, CA, USA) are 0.4 kV and 125 μF. The time constant is usually approximately 1.6. Record the actual voltage, time constant, and any other relevant parameters after electroporation. This provides a useful starting point for troubleshooting if problems are encountered.

17. Incubate the cell suspension in the transfection cuvette for 10 minutes at room temperature. During this time feed the MEF with 10 mL of 37°C medium/LIF. Do not add selection agents (e.g., G418, FIAU, hygromycin, etc.) at this time.

18. Mix the cell suspension in the cuvette using a 2-mL pipet and distribute evenly between the five 100-mm plates of MEF.

19. Transfer the plates to a CO_2 incubator.

20. Add selection medium and pick colonies as described in Table 2 and in the subsequent instructions.

The schedule for transfecting, selecting, picking, expanding, and preserving drug-resistant clones will vary between different ES cell lines and even between different targeting constructs introduced into the same cell line. One must carefully judge the rate of cell killing, the survival of drug-resistant clones, and the expansion of clones once picked. Monitoring the same criteria of cell and colony morphology used above for derivation and maintenance of ES cell lines is also critical during the selection process. Even if a culture contains only a few isolated colonies, these cells should be passaged, without expansion of the culture as a whole if necessary, based on the appearance of individual colonies. It is essential to expand cells by introducing additional culture passages rather than by allowing colonies to overgrow and differentiate.

We present a protocol in which expansion of ES cell clones and DNA isolation are performed using 24-well plates and in which cells are cryopreserved prior to genotype analysis. Alternatively, one can use smaller culture geometries (96-well plates) and perform genotype analyses while the ES cell clones remain in culture without cryopreservation. This can be much more efficient, because one can cryopreserve only the targeted clones and avoid the labor of preserving unwanted nontargeted clones. However, careful timing and rapid (by PCR) reliable genotype analysis is essential. This approach is thus best implemented after one has acquired significant experience performing homologous recombination in ES cells.

When picking ES cell colonies, the objective is to pick individual colonies, fully dissociate them, and transfer the dispersed cells to individual culture wells as efficiently as possible. Colonies should be picked when they contain between one hundred and several hundred cells, but before morphologic evidence of differentiation is observed, as described above. Individual colonies will attain these criteria at different times after transfection, and colonies can therefore be picked from any given plate during a several day period.

1. Rinse one 100-mm plate of transfected ES cells twice with 2 mL HBS.

2. Add 5 mL HBS to this plate.

3. Aliquot 10 µL HBS into each well of V bottom 96-well plates.

4. Use sterile round sequencing pipet tips to pick and transfer individual colonies to the HBS in the separate microwells. This is conveniently done in cohorts of 8 or fewer colonies.

5. Add 70 µL Trypsin/EDTA/HBS to each microwell containing a

nondissociated colony and incubate this plate for 3 minutes at 37°C in 5% CO_2.

6. Add 70 µL complete medium/LIF to each microwell containing ES cells from 5.

7. Triturate all 8 wells at once using a multichannel pipet set at 150 µl.

8. Transfer each dissociated colony to MEF in a C24 well containing 1 mL medium/LIF/selection agents.

9. Repeat the above process as necessary. When all colonies have been picked, add 10 mL medium/LIF/selection agents to the 100-mm plate, and return it to the incubator.

10. Repeat the above process for each plate of transfected ES cells.

11. Additional colonies may become visible after additional culture of plates from which early colonies have already been picked. These later colonies may also be picked as long as they do not represent secondary, or satellite, colonies seeded from primary colonies.

12. Feed C24 plates daily with 1 mL medium/LIF/selection agents per well until well-formed colonies appear without signs of differentiation.

The cultures used to isolate DNA can overgrow and differentiate significantly, as long as the cells remain viable, without compromising the quality of DNA obtained or one's ability to genotype the clones.

1. Rinse each C24 well containing individual ES cell clones twice with 0.5 mL HBS.

2. Add 3 drops Trypsin/EDTA/HBS using a 2-mL pipet.

3. Incubate for 3 minutes at 37°C in 5% CO_2.

4. Add 1 mL complete medium without LIF or selection drugs.

5. Triturate well with a Pasteur pipet and examine the result using a phase-contrast microscope. If the majority of cells are not single, then repeat the trituration.

6. Add all except 0.5 mL of these cells (approximately 0.6 mL total is transferred) to a gelatinized C24 well containing 0.5 mL medium/LIF but without additional MEF. These samples will be cultured for isolation of DNA.

7. Add the reserved 0.5 mL single cell suspension to a cryovial on ice/water containing 0.5 mL cold 20% DMSO/complete medium (without LIF or selection agents). Additional serum in the freezing medium may enhance recovery of cells but is not absolutely necessary. Mix well, freeze using a programmed LN2 cell freezer or by placing the vials overnight in an insulated container at -80°C, and transfer to LN2.

Table 2. Schedule For Transfecting, Selecting, Expanding, and Preserving Genetically Modified ES Cells

Day of Experiment	Procedure
Day 1	Plate MEF in 100-mm plate.
Day 2	Plate ES cells on MEF in 100-mm plate.
Day 3	Plate MEF in five 100-mm plates.
Day 4	Transfect 1×10^7 ES cells and plate these on five 100-mm plates of MEF.
Day 5	Start G418 selection (0.2 mg active drug/mL) 20 to 21 hours after transfection.
Day 6	Add FIAU or gancyclovir 48 hours after transfection.
Day 6 to 13	Feed daily with medium/LIF/G418/FIAU. Supplement feeder cells to maintain a confluent but not overcrowded monolayer.
Day 10	Plate MEF in C24 wells.
Day 11 to 13	Pick colonies (see text) and plate on MEF in C24 wells with medium/LIF/G418 (without FIAU).
Day 11 to 17	Feed C24 plates daily with medium/LIF/G418 (without FIAU). Continue feeding primary 100-mm plates with medium/LIF/G418/FIAU until all colonies have been picked.
Day 15 to 17	Trypsinize, freeze, and expand clones for DNA preparation (see text).

8. Feed the remaining C24 wells daily with medium/LIF until the cells approach confluence.

9. Rinse confluent wells with HBS, add 250 µL proteinase K digestion buffer, and return the plates to the incubator until all wells have been digested for at least 12 hours. This incubation with proteinase K can be continued for at least 7 days if the volume in the wells is maintained by adding water. After all wells are digested, store the plates at -20°C. This freezing step may aid later solubilization of high molecular weight DNA (see below).

To identify homologous recombinant ES cell clones, the most reliable method is Southern blot analysis.

Appropriate restriction enzymes and a probe that is located outside sequences within the targeting construct must be carefully characterized using genomic DNA. Typically, a probe from either 5′ or 3′ of the targeting construct is used for screening, and a second probe from the other end is used for verification. Southern analysis for *neo*, or other selectable marker sequences, should also be performed to verify that ES cell clones carry only a single insertion site of transfected DNA.

The protocol for lysing cells directly in 24-well plates has been described previously (29) and is convenient, but produces extremely high molecular weight DNA. In our experience, this DNA is difficult to solubilize by several standard techniques, and certain isolation protocols have resulted in irretrievable loss of useful DNA. However, vortex mixing at room temperature and heating at 65°C overnight, as described below, is easy and uniformly successful.

1. Add 0.5 mL of 100% EtOH to each C24 well containing 250 µL proteinase K digests of ES cell clones. Swirl the plate vigorously. DNA should form an obvious spool in the center of the well. If not, the DNA may be scraped from the bottom of the well using a plastic micropipet tip.

2. Using a fresh micropipet tip for each sample, lift the spool of DNA out of the well and dunk it 4 to 5 times in absolute EtOH followed by 4 to 5 times in 70% EtOH. Dry momentarily before placing the DNA into a microcentrifuge tube containing approximately 100 µL 10 mM Tris-HCl, pH 7.6, 1 mM EDTA (TE). Do not over-dry! The volume of TE should be adjusted to the relative size of the DNA spool. A constant volume of DNA solution can then be used in restriction enzyme digests, and the load will be more consistent without requiring quantitation of every sample.

3. Vortex mix each sample at maximum speed for 5 minutes.

4. Heat samples overnight at 65°C.

5. Quick-spin all samples, and ensure that each is fluid-like, and that all DNA spools are dissolved.

6. Use 16 µL of the DNA samples in 20 µL restriction enzyme reactions for Southern blot analysis using a 20-cm-wide agarose gel and 36-tooth comb. Add the enzyme in two equal doses at the beginning and half way through a 4-hour reaction period. Additional bovine serum albumin may be added (0.1 mg/mL) to counteract detrimental effects of residual SDS. Alternatively, use 1 µL of the DNA samples in 20 µL PCRs.

Introducing ES Cells into Embryos

In this section, we will outline the general procedures and provide references for the two major methods currently used to generate ES cell embryo chimeras. Historically, the first approach was injection of ES cells into blastocysts (48). More recently, aggregation of ES cells with morulae has been developed (40,63). It cannot be overemphasized, however, that success using either method (meaning the contribution of mutant ES cells to the mouse germline) is absolutely dependent on the totipotency of the ES cells as well as the health of the recipient embryos.

Blastocyst injection has been the most widely used method and involves injection of 10 to 15 ES cells into the blastocoel cavity of preimplantation embryos. After a short time in culture, the injected embryos are surgically transferred to the uterus of a pseudopregnant recipient. While this procedure has yielded wide success in generation of germline chimeras, it has some legitimate practical limitations. The major drawbacks are the cost of the equipment required and the time and effort required to train personnel to perform blastocyst injection. Specifically, the equipment required for such delicate handling of blastocysts and ES cells, such as holding pipets and injection needles, must be made by the investigator. Also, the precise qualities of these materials often dictate the success rate. Moreover, the ability to perform successful blastocyst injections, in a timely manner, is not easily acquired. For these reasons, aggregation of ES cells with embryos has been investigated as an alternative method of generating chimeras.

An informative and practical comparison of the two procedures (62) highlights the successes of ES cell aggregation with morulae. As mentioned, sophisticated equipment is not required. In addition, between 20 and 30 morula stage embryos can be obtained per donor mouse. This is because morula stage embryos can be obtained from each mouse by superovulation, compared to the average of 6 to 8 blastocysts obtained per mouse by natural mating. After relatively simple removal of the zona pellucida, the ES cells and morulae are brought into close physical contact and aggregate spontaneously. ES cells initially attach to the outside of the morula, and after overnight culture, the vast majority of embryos have incorporated many ES cells into the ICM. During this culture period, the morulae have also progressed to blastocyst stage and are then surgically transferred to the uterus of a pseudopregnant recipient.

Also, included in this comparison is data from a systematic analysis using several independent clones of the early passage pluripotent ES cell line known as R1 (41). The authors compared, for both blastocyst injection and aggregation, the number of embryos transferred, frequency of chimeras obtained, and frequency of germline transmission. While frequency of chimerism appeared higher with blastocyst injection, the overall frequency of germline transmission was the same for the two techniques. As germline transmission is the ultimate goal in such experiments, it is the most important parameter to examine.

Thus, ES cell morula aggregation appears to be a very promising alternative approach. Given the time and effort already involved in this type of in vivo analysis, it would be wise to seriously consider aggregation as the method of choice. This seems especially relevant for investigators new to the field of gene targeting.

Deriving Homozygous Mutant ES Cells via Loss of Heterozygosity: Implications for Sequential Gene Targeting Strategies

Heterozygous ES cells maintained in culture frequently produce clones lacking one of the two alleles originally present at the mutant locus (38). This LOH proceeds in both directions, resulting in some clones homozygous for each of the two original alleles. Cell lines homozygous for alleles containing selectable markers can be obtained efficiently by several approaches (see below). Coupled with chimera analysis, LOH can be a useful adjunct to germline transmission if either homozygous (2,50,57,64) or heterozygous (7) mutant conceptuses do not survive. Chimeras allow one to study cells of different genotypes in an essentially competitive environment and may reveal functions not observed in nonchimeric mutant animals (11,50).

LOH has also been combined with in vitro differentiation of ES cells to analyze the effect of targeted null mutations on cardiocyte function, hematopoiesis, and adipogenesis (50,54,55). In vitro differentiation mimics many of the complex developmental lineage relationships of normal embryogenesis and also produces useful, physiologically important, differentiated cell types. Using this approach, a greater number of mutant alleles can be analyzed than would be practical via germline transmission. One can also more readily introduce wild-type and mutant rescue constructs to test structure–function hypotheses.

LOH also has practical implications for experimental design. During multistep mutation strategies using negative selection in a second, or later, step, LOH producing homozygous wild-type cells introduces a background of clones that are resistant to negative selection, but lack the desired mutation that is usually intended to delete the negative selectable marker. These clones are easily identified by DNA analysis, but can significantly increase the labor required to identify clones carrying the desired second mutation. In addition, LOH probably also occurs at loci undergoing spontaneous mutation during ES cell culture. These homozygous mutations may have unanticipated effects when introduced into embryos.

The mechanism of LOH in ES cells has not been established, but may involve gene conversion or chromosomal loss coupled with duplication of the remaining chromosome. The latter process would lead to homozygosity of a larger number of genetically linked alleles. Clones homozygous for alleles bearing positive selectable markers, such as *neo*, can be enriched by culture of heterozygotes with high concentrations of selection agent (38). This approach can be applied to two or more loci in the same cell line by using different selectable markers, conferring resistance to G418, hygromycin, puromycin, etc. (37), to target different loci or, more efficiently, by recycling the *neo* marker using Cre recombinase (35). An alternative method of producing homozygous mutations utilizes two sequential targeting steps with different selectable markers to modify both wild-type alleles (39,58).

CONDITIONAL AND INTRA-GENE TARGETING

Conditional gene targeting refers to cell-type-specific and/or inducible gene targeting. The most common reason for taking this approach is to avoid or bypass an embryonic lethal phenotype of a systemic mutation. Also, because genes frequently act in several cell types, the complex phenotype of a whole-animal mutation may be difficult or impossible to interpret. Intra-gene targeting refers to structurally modifying an endogenous gene to produce an altered gene product. One example of this approach is modification or deletion of an alternatively spliced exon. In both conditional and intra-gene targeting, the modified gene must be expressed normally at some point, either during development and/or in the adult. Thus, the precise nature of the mutation introduced is critical.

While at the current time there is no guaranteed method of creating conditional mutations in the mouse, the most successful system so far utilizes the site-specific Cre–loxP recombinase (53). Several successes, as well as failures, are described in a recent meeting review that is also available online at **http://www.leadingstrand.org** (51). The Cre recombinase can remove virtually any DNA sequence located between two direct repeat copies of the 34-bp loxP recognition sequence (23) and has also been used to create large deletions and inversions (47). This recombinase is active in both mammalian cells in culture and in mice. This has allowed Rajewsky and colleagues to develop a strategy in which conditional mutations are made in endogenous genes in ES cells (46). Generation of intra-gene mutations is accomplished by similar methods.

Obtaining Intra-Gene Mutations with the Cre–LoxP Recombinase System

In this approach, both the target gene, as well as a selectable marker expression cassette, are flanked by loxP sites in the targeting vector. Figure 2 shows a typical vector design, which requires two transfections in ES cells: the first achieves stable incorporation of the targeting construct, and the second removes the selection cassette via Cre–loxP-mediated recombination. This is required because the most common selections can interfere with expression of the targeted gene by introducing cryptic transcription initiation sites, or RNA splice signals, and by unknown mechanisms (18).

In the first transfection, classical gene targeting procedures are used with the aid of a positive selection marker such as *neo* and selection in G418. The procedures involved are the same as those described earlier in this chapter for producing heterozygous targeted ES cell clones, except that the negative selection agent may be omitted.

Following identification of targeted clones, further molecular analysis of the mutant allele is suggested, to determine if any clones represent incomplete targeting events. This refers to homologous recombination between the endogenous gene and the genomic region in the targeting vector, which is flanked by loxP sites (stippled exon in Figure 2). If such recombination occurs, then the most distal loxP site will be lost, rendering the clone useless for further modification. In our experience, the frequency of this event appears to be significant. In one example, 4 of 17 targeted clones carried such an insertion. Others have observed that up to 50% of targeted clones can lose the most distal loxP site (21,33).

To achieve Cre–loxP-mediated recombination in ES cells, a supercoiled Cre recombinase expression plasmid is transfected by electroporation. Use of a supercoiled plasmid prevents integration into the genome. Many such plasmids are available, and all should be tested for functional recombinase activity using previously characterized cells. This testing is necessary only because some routine plasmid preparations have resulted in mutations that disrupt expression, nuclear localization, or recombination.

Ideally, the previous step of classical gene targeting has gone well, and several targeted clones are now available for Cre–loxP-mediated recombination. As this process requires further culture and selection, in which totipotency of ES cells may be lost, it is best to select several targeted clones for this step. Attempts should be made to select as many clones as possible that consistently display optimal growth characteristics. We have also found that keeping a brief history of each targeted clone is helpful. Generation of a grid, matching the 24-well plates in which picked clones are cultured, helps facilitate this task. Such a brief history could include the following: morphology and size of each clone at initial picking, rate of growth once dispersed, and morphology of subsequent colonies at freezing. If enough independent targeted clones are available, we have found that selecting five for the Cre–loxP-mediated recombination step works well, and is quite manageable. Many of the procedures are the same as those described earlier in this chapter.

1. Prepare enough supercoiled Cre recombinase expression plasmid to use 20 µg per transfection.

2. Prepare approximately 5×10^6 cells, from each targeted clone, for electroporation as described earlier.

3. Combine 20 µg plasmid DNA and suspended cells in a single electroporation cuvette and electroporate as described earlier.

4. Plate all cells from each transfection onto one 10-cm dish with MEF feeders and no selection drugs.

5. At 72 hours posttransfection, add a negative selection drug if desired (see below). This delay in addition is crucial, in order to allow enough time for expression of Cre and homologous recombination, followed by degradation of cytoplasmic thymidine kinase activity.

6. At day 4 or 5 after selection begins, pick viable clones. This is considerably earlier in selection than with many other drugs, and viable colonies are less easily distinguished from dying colonies. However, use of low-power magnification (4× objective, with 10× eye pieces) without phase-contrast facilitates this. Under these conditions, viable colonies have a sharp dark line around the periphery, and the entire interior of the colony is bright white compared to the feeder layer. Actual picking and expansion procedures are the same as those described earlier in this chapter.

7. Perform Southern blot analysis of genomic DNA, using appropriate restriction enzymes and probes, to distinguish between type I and type II deletions (See Figure 2).

Frequencies of gene targeting are known to vary depending on the locus to be targeted (17). In addition, we have observed that the same locus may be targeted with widely varying frequency in different ES cell lines. The following frequencies are results of experiments with a single targeting vector. In D3 ES cells (13), only 2 of 672 (1 of 336) transfected clones were targeted. In contrast, J1 ES cells (32) yielded approximately 10-fold more targeted clones (14 of 432 [1 of 31] transfectants). If problems with initial targeting are encountered, these data suggest that an alternative ES cell line might be more successful.

In contrast, we have observed virtually no differences in Cre–loxP-mediated recombination frequency in these two ES cell lines. In both lines, virtually all gancyclovir-resistant clones harbor either a type I or type II deletion. The type I deletion, in both cell lines, occurs about 5-fold more often than the type II deletion. This difference in frequency has also been observed in several other laboratories, and attempts have been made to increase the frequency of type II deletions by reducing the amount of functional Cre recombinase available (33).

As an alternative approach, we have also investigated the frequency of Cre–loxP-mediated recombination without the use of a negative selection scheme. In our hands, the frequency of Cre–loxP-mediated recombination is so high that selection does not appear absolutely necessary. This is in stark

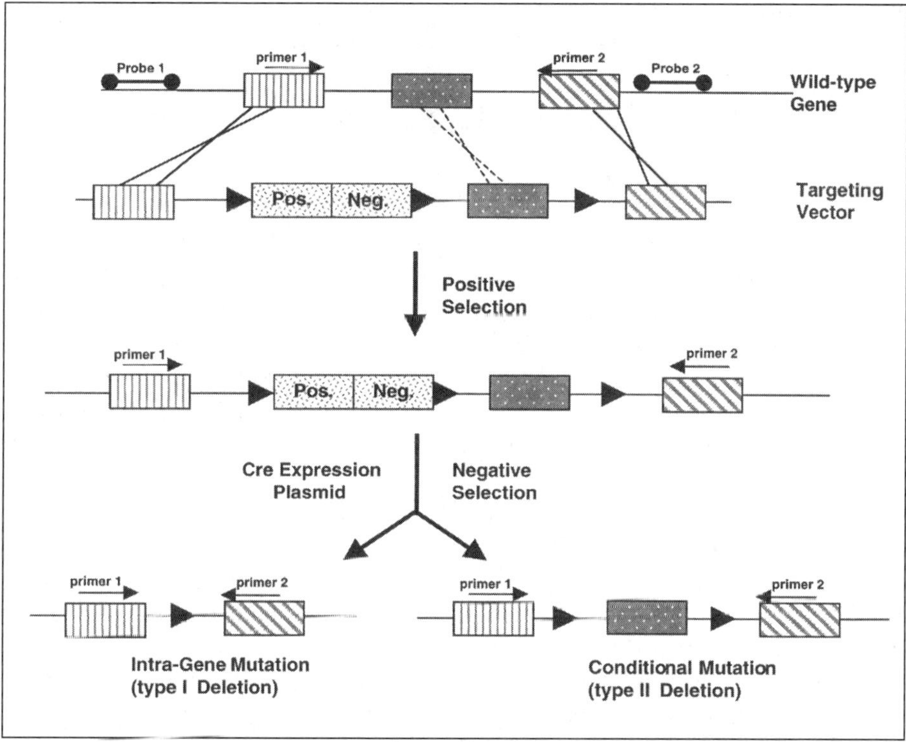

Figure 2. Vector design for conditional and intra-gene targeting. In this approach, two insertions are made in the wild-type gene: a positive and negative selection cassette, which is surrounded by loxP sites (black triangles), and a single loxP site. For conditional knockout mutations, the stippled box represents the first coding exon. For intra-gene targeting, the dotted box represents any exon of interest. For initial screening of ES clones, Southern blot analysis is recommended over PCR, because a deletion has not been made in the targeting vector. Two genomic probes, located on opposite sides of the plasmid mutation, are required for screening, and only the first must lie outside of the targeting vector sequence. The second probe should be from the opposite end of the targeting construct, near the single loxP site. The purpose of this probe is to determine which targeted cells have undergone homologous recombination between loxP sites (dashed lines), resulting in loss of the flanking loxP site. PCR screening is useful after Cre–loxP-mediated recombination, and one set of two primers can be used to assay the wild-type allele as well as both mutant alleles (type I and type II deletions). Horizontal lines represent introns and flanking sequences. Boxes represent exons (striped and dotted patterns) or selection genes (stippled pattern). Horizontal arrows represent PCR primers and their direction of extension. Probes outside the targeting vector, at both ends, are indicated.

contrast to the first (gene targeting) step, which absolutely requires at least a positive selection marker. In our hands, all five independent clones, assayed efficiently, produced mutant cells bearing type I or type II deletions after transfection with Cre, without negative selection. Actual frequencies are as follows: 12/102 (12%), 5/90 (6%), 4/94 (4%), 60/78 (77%), and 16/94 (17%). The procedural modification is very straightforward; simply omit selection drugs after transfection with a Cre-expressing plasmid. On the seventh day after transfection, pick colonies using the same criteria as for a selection approach. It must be emphasized however, that while our results are very encouraging, they may not be typical for other loci.

Obtaining Conditional Mutations by Breeding With Cre-Expressing Transgenic Mice

Conditional gene targeting of a loxP-flanked (floxed) gene or gene segment requires expression of Cre recombinase in a cell-type- and/or stage-specific manner. This is most simply achieved by breeding the floxed mutant strain with a Cre-expressing transgenic strain. This approach was first accomplished by collaboration between the K. Rajewsky and J. Marth groups, using the *lck-cre* transgene to make a cell-type-specific deletion in T cells (19). In principle, a single floxed mutant strain can be used in conjunction with several different Cre-expressing transgenic strains. This allows investigation of the specific functions of a particular gene in a variety of cell types in vivo, without the confusion inherent in whole animal mutations.

Attempts to generate useful Cre-expressing transgenic strains have met with many successes as well as failures. Often, low expression levels of Cre recombinase, resulting in mosaic excision accompany the desired tissue specificity. To allow detection of such problems, reporter mouse lines have been developed that indicate Cre activity. In these lines, Cre excision results in activation of a ubiquitously expressed cell marker (51). No single marker has been found that works in all tissues, but most tissues can be assessed by use of one of the lines developed.

Some of the most recent examples of success in conditional mutagenesis using the Cre–loxP system have yielded interesting and useful results in a variety of fields of study. Use of the rat insulin promoter to drive Cre recombinase allowed generation of a pancreatic β cell-specific knockout of the insulin receptor gene (28). The mutant mice are defective in insulin secretion in response to glucose, an effect that was not anticipated, but is similar to the behavior of β cells in type 2 diabetes. In the cardiovascular field, ventricle-specific Cre has been used to make a restricted deletion of gp130 (22). These mutant mice have revealed a critical role for gp130-dependent cardiomyocyte survival in the transition to heart failure during biomechanical stress. In a collaboration between the fields of developmental fate specification and immune response, an interferon-regulated Cre has allowed generation of a *Notch1*-deficient bone marrow. These mutant mice displayed a complete block in development of the T cell lineage (45).

The likelihood of success in obtaining the desired cell-type-specific or inducible Cre-expressing transgenic strain is improving every day. Importantly, these advances are due not only to experience in individual laboratories, but also to the cooperation between laboratories in sharing useful strains of mice. At present, an organized system for information exchange in such matters is in its infancy. One private database of Cre-expressing strains exists and is maintained by Dr. Andras Nagy (**http://www.mshri.on.ca/nagy/cre.htm**). Clearly, given the recent advances in manipulating the mouse genome, and the additional, more powerful, tools still being developed, we should expect many more contributions, to both the database and to our overall understanding, in the near future.

ACKNOWLEDGMENTS

We thank Drs. Richard Mortensen, Rudolf Jaenisch, En Li, and Arlene Sharpe for invaluable advice over several years regarding derivation, maintenance, and genetic manipulation of ES cells and blastocyst-mediated transgenesis. Work in the authors' laboratories is supported by National Institutes of Health Grant Nos. GM57719 and GM52366 (to/ E.L.G.) and HL54095 and HL30628 (to D.S.M.)

REFERENCES

1. Adra, C.N., P.H. Boer, and M.W. McBurney. 1987. Cloning and expression of the mouse pgk-1 gene and the nucleotide sequence of its promoter. Gene *60*:65-74.
2. Arroyo, A.G., J.T. Yang, H. Rayburn, and R.O. Hynes. 1996. Differential requirements for alpha4 integrins during fetal and adult hematopoiesis. Cell *85*:997-1008.
3. Bader, B.L., H. Rayburn, D. Crowley, and R.O. Hynes. 1998. Extensive vasculogenesis, angiogenesis, and organogenesis precede lethality in mice lacking all alpha v integrins. Cell *95*:507-519.
4. Baldwin, H.S., C. Mickanin, and C. Buck. 1997. Adenovirus-mediated gene transfer during initial organogenesis in the mammalian embryo is promoter-dependent and tissue-specific. Gene Ther. *4*:1142-1149.
5. Bishop, J.O. and P. Smith. 1989. Mechanism of chromosomal integration of microinjected DNA. Mol. Biol. Med. *6*:283-298.
6. Brooks, P.C., R.A. Clark, and D.A. Cheresh. 1994. Requirement of vascular integrin alpha v beta 3 for angiogenesis. Science *264*:569-571.
7. Carmeliet, P., V. Ferreira, G. Breier, S. Pollefeyt, L. Kieckens, M. Gertsenstein, M. Fahrig, A. Vandenhoeck, K. Harpal, C. Eberhardt, et al. 1996. Abnormal blood vessel development and lethality in embryos lacking a single VEGF allele. Nature *380*:435-439.
8. Cepko, C.L., S. Fields-Berry, E. Ryder, C. Austin, and J. Golden. 1998. Lineage analysis using retroviral vectors. Curr. Top. Dev. Biol. *36*:51-74.

155

9. Choi, T., M. Huang, C. Gorman, and R. Jaenisch. 1991. A generic intron increases gene expression in transgenic mice. Mol. Cell. Biol. *11*:3070-3074.

10. Copp, A.J. 1995. Death before birth: clues from gene knockouts and mutations [see comments]. Trends Genet. *11*:87-93.

11. Crosby, J.R., R.A. Seifert, P. Soriano, and D.F. Bowen-Pope. 1998. Chimaeric analysis reveals role of Pdgf receptors in all muscle lineages. Nat. Genet. *18*:385-388.

12. Deng, C. and M.R. Capecchi. 1992. Reexamination of gene targeting frequency as a function of the extent of homology between the targeting vector and the target locus. Mol. Cell. Biol. *12*:3365-3371.

13. Doetschman, T.C., H. Eistetter, M. Katz, W. Schmidt, and R. Kemler. 1985. The in vitro development of blastocyst-derived embryonic stem cell lines: formation of visceral yolk sac, blood islands and myocardium. J. Embryol. Exp. Morphol. *87*:27-45.

14. Frenette, P.S., T.N. Mayadas, H. Rayburn, R.O. Hynes, and D.D. Wagner. 1996. Susceptibility to infection and altered hematopoiesis in mice deficient in both P- and E-selectins. Cell *84*:563-574.

15. Gardner, D.P., G.W. Byrne, F.H. Ruddle, and C. Kappen. 1996. Spatial and temporal regulation of a lacZ reporter transgene in a binary transgenic mouse system. Transgenic Res. *5*:37-48.

16. George, E.L., E.N. Georges-Labouesse, R.S. Patel-King, H. Rayburn, and R.O. Hynes. 1993. Defects in mesoderm, neural tube and vascular development in mouse embryos lacking fibronectin. Development *119*:1079-1091.

17. George, E.L. and R.O. Hynes. 1994. Gene targeting and generation of mutant mice for studies of cell-extracellular matrix interactions. Methods Enzymol. *245*:386-420.

18. Georges-Labouesse, E.N., E.L. George, H. Rayburn, and R.O. Hynes. 1996. Mesodermal development in mouse embryos mutant for fibronectin. Dev. Dyn. *207*:145-156.

19. Gu, H., J.D. Marth, P.C. Orban, H. Mossmann, and K. Rajewsky. 1994. Deletion of a DNA polymerase beta gene segment in T cells using cell type-specific gene targeting [see comments]. Science *265*:103-106.

20. Hanahan, D. 1989. Transgenic mice as probes into complex systems. Science *246*:1265-1275.

21. Hennet, T., F.K. Hagen, L.A. Tabak, and J.D. Marth. 1995. T-cell-specific deletion of a polypeptide N-acetylgalactosaminyl-transferase gene by site-directed recombination. Proc. Natl. Acad. Sci. USA. *92*:12070-12074.

22. Hirota, H., J. Chen, U.A. Betz, K. Rajewsky, Y. Gu, J. Ross, Jr., W. Muller, and K.R. Chien. 1999. Loss of a gp130 cardiac muscle cell survival pathway is a critical event in the onset of heart failure during biomechanical stress. Cell *97*:189-198.

23. Hoess, R.H., M. Ziese, and N. Sternberg. 1982. P1 site-specific recombination: nucleotide sequence of the recombining sites. Proc. Natl. Acad. Sci. USA. *79*:3398-3402.

24. Hogan, B., R. Beddington, F. Costantini, and E. Lacy. 1994. Manipulating the Mouse Embryo: A Laboratory Manual, 2nd ed. CSH Laboratory Press, Cold Spring Harbor, NY.

25. Hynes, R.O. 1996. Targeted mutations in cell adhesion genes: what have we learned from them? Dev. Biol. *180*:402-412.

26. Jaenisch, R. 1988. Transgenic animals. Science *240*:1468-1474.

27. Jarnagin, W.R., D.C. Rockey, V.E. Koteliansky, S.S. Wang, and D.M. Bissell. 1994. Expression of variant fibronectins in wound healing: cellular source and biological activity of the EIIIA segment in rat hepatic fibrogenesis. J. Cell Biol. *127*:2037-2048.

28. Kulkarni, R.N., J.C. Bruning, J.N. Winnay, C. Postic, M.A. Magnuson, and C.R. Kahn. 1999. Tissue-specific knockout of the insulin receptor in pancreatic beta cells creates an insulin secretory defect similar to that in type 2 diabetes. Cell *96*:329-339.

29. Laird, P.W., A. Zijderveld, K. Linders, M.A. Rudnicki, R. Jaenisch, and A. Berns. 1991. Simplified mammalian DNA isolation procedure. Nucleic Acids Res. *19*:4293.

30. Lawler, J., M. Sunday, V. Thibert, M. Duquette, E.L. George, H. Rayburn, and R.O. Hynes. 1998. Thrombospondin-1 is required for normal murine pulmonary homeostasis and its absence causes pneumonia. J. Clin. Invest. *101*:982-992.

31. Leconte, I., J.C. Fox, H.S. Baldwin, C.A. Buck, and J.L. Swain. 1998. Adenoviral-mediated expression of antisense RNA to fibroblast growth factors disrupts murine vascular development. Dev. Dyn. *213*:421-430.

32. Li, E., T.H. Bestor, and R. Jaenisch. 1992. Targeted mutation of the DNA methyltransferase gene results in embryonic lethality. Cell *69*:915-926.

33. Marth, J.D. 1996. Recent advances in gene mutagenesis by site-directed recombination. J. Clin. In-

vest. *97*:1999-2002.

34. Mayadas, T.N., R.C. Johnson, H. Rayburn, R.O. Hynes, and D.D. Wagner. 1993. Leukocyte rolling and extravasation are severely compromised in P selectin-deficient mice. Cell *74*:541-554.

35. Milstone, D.S., G. Bradwin, and R.M. Mortensen. 1999. Simultaneous Cre catalyzed recombination of two alleles to restore neomycin sensitivity and facilitate homozygous mutations. Nucleic Acids Res. *27*:e10.

36. Milstone, D.S., D. Fukumura, R.C. Padgett, P.E. O'Donnell, V.M. Davis, O.J. Benavidez, W.L. Monsky, R.J. Melder, R.K. Jain, and M.A. Gimbrone, Jr. 1998. Mice lacking E-selectin show normal numbers of rolling leukocytes but reduced leukocyte stable arrest on cytokine-activated microvascular endothelium. Microcirculation *5*:153-171.

37. Mortensen, R.M. 1998. Gene targeting by homologous recombination, p. 9.15-9.17. *In* F.M. Ausubel, R. Brent, R.E. Kingston, D.D. Moore, J.G. Seidman, J.A. Smith, and K. Struhl (Eds.), Current Protocols in Molecular Biology. John Wiley & Sons, Philadelphia, PA.

38. Mortensen, R.M., D.A. Conner, S. Chao, A.A. Geisterfer-Lowrance, and J.G. Seidman. 1992. Production of homozygous mutant ES cells with a single targeting construct. Mol. Cell. Biol. *12*:2391-2395.

39. Mortensen, R.M., M. Zubiaur, E.J. Neer, and J.G. Seidman. 1991. Embryonic stem cells lacking a functional inhibitory G-protein subunit (alpha i2) produced by gene targeting of both alleles. Proc. Natl. Acad. Sci. USA *88*:7036-7040.

40. Nagy, A., E. Gocza, E.M. Diaz, V.R. Prideaux, E. Ivanyi, M. Markkula, and J. Rossant. 1990. Embryonic stem cells alone are able to support fetal development in the mouse. Development *110*:815-821.

41. Nagy, A., J. Rossant, R. Nagy, W. Abramow-Newerly, and J.C. Roder. 1993. Derivation of completely cell culture derived mice from early-passage embryonic stem cells. Proc. Natl. Acad. Sci. USA *90*:8424-8428.

42. Ornitz, D.M., R.W. Moreadith, and P. Leder. 1991. Binary system for regulating transgene expression in mice: targeting int-2 gene expression with yeast GAL4/UAS control elements. Proc. Natl. Acad. Sci. USA *88*:698-702.

43. Palmiter, R.D., and R.L. Brinster. 1986. Germ-line transformation of mice. Annu. Rev. Genet. *20*:465-499.

44. Palmiter, R.D., E.P. Sandgren, M.R. Avarbock, D.D. Allen, and R.L. Brinster. 1991. Heterologous introns can enhance expression of transgenes in mice. Proc. Natl. Acad. Sci. USA *88*:478-482.

45. Radtke, F., A. Wilson, G. Stark, M. Bauer, J. van Meerwijk, H.R. MacDonald, and M. Aguet. 1999. Deficient T cell fate specification in mice with an induced inactivation of Notch1. Immunity *10*:547-558.

46. Rajewsky, K., H. Gu, R. Kuhn, U.A. Betz, W. Muller, J. Roes, and F. Schwenk. 1996. Conditional gene targeting. J. Clin. Invest. *98*:600-603.

47. Ramirez-Solis, R., P. Liu, and A. Bradley. 1995. Chromosome engineering in mice. Nature *378*:720-724.

48. Robertson, E.J. (Ed.). 1987. Teratocarcinomas and Embryonic Stem Cells: A Practical Approach. IRL Press, Oxford.

49. Robinson, S.D., P.S. Frenette, H. Rayburn, M. Cummiskey, M. Ullman-Cullere, D.D. Wagner, and R.O. Hynes. 1999. Multiple, targeted deficiencies in selectins reveal a predominant role for P-selectin in leukocyte recruitment. Proc. Natl. Acad. Sci. USA *96*:11452-11457.

50. Rosen, E.D., P. Sarraf, A.E. Troy, G. Bradwin, K. Moore, D.S. Milstone, B.M. Spiegelman, and R.M. Mortensen. 1999. PPAR gamma is required for the differentiation of adipose tissue in vivo and in vitro. Mol. Cell *4*:611-617.

51. Rossant, J. and A. McMahon. 1999. "Cre"-ating mouse mutants—a meeting review on conditional mouse genetics. Genes Dev. *13*:142-145.

52. Rudnicki, M.A., T. Braun, S. Hinuma, and R. Jaenisch. 1992. Inactivation of MyoD in mice leads to up-regulation of the myogenic HLH gene Myf-5 and results in apparently normal muscle development. Cell *71*:383-390.

53. Sauer, B. and N. Henderson. 1988. Site-specific DNA recombination in mammalian cells by the Cre recombinase of bacteriophage P1. Proc. Natl. Acad. Sci. USA. *85*:5166-5170.

54. Simon, M.C., L. Pevny, M.V. Wiles, G. Keller, F. Costantini, and S.H. Orkin. 1992. Rescue of erythroid development in gene targeted GATA-1- mouse embryonic stem cells. Nat. Genet. *1*:92-98.

55.Sowell, M.O., C.P. Ye, D.A. Ricupero, S. Hansen, S.J. Quinn, P.M. Vassilev, and R.M. Mortensen. 1997. Targeted inactivation of alpha(I2) or alpha(I3) disrupts activation of the cardiac muscarinic K+ channel, Ik+Ach, in intact cells. Proc. Natl. Acad. Sci. USA *94*:7921-7926.

56.Spencer, D.M., P.J. Belshaw, L. Chen, S.N. Ho, F. Randazzo, G.R. Crabtree, and S.L. Schreiber. 1996. Functional analysis of Fas signaling in vivo using synthetic inducers of dimerization. Curr. Biol. *6*:839-847.

57.Taverna, D., M.H. Disatnik, H. Rayburn, R.T. Bronson, J. Yang, T.A. Rando, and R.O. Hynes. 1998. Dystrophic muscle in mice chimeric for expression of alpha5 integrin. J. Cell Biol. *143*:849-859.

58.te Riele, H., E.R. Maandag, A. Clarke, M. Hooper, and A. Berns. 1990. Consecutive inactivation of both alleles of the pim-1 proto-oncogene by homologous recombination in embryonic stem cells. Nature *348*:649-651.

59.Thomas, K.R. and M.R. Capecchi. 1987. Site-directed mutagenesis by gene targeting in mouse embryo-derived stem cells. Cell *51*:503-512.

60.Tucker, K.L., Y. Wang, J. Dausman, and R. Jaenisch. 1997. A transgenic mouse strain expressing four drug-selectable marker genes. Nucleic Acids Res. *25*:3745-3746.

61.Wasserman, P.M. and M.L. DePamphilis (Ed.). 1993. Guide to Techniques in Mouse Development, Vol. 225. Academic Press, Boston.

62.Wood, S.A., N.D. Allen, J. Rossant, A. Auerbach, and A. Nagy. 1993. Non-injection methods for the production of embryonic stem cell-embryo chimaeras. Nature *365*:87-89.

63.Wood, S.A., W.S. Pascoe, C. Schmidt, R. Kemler, M.J. Evans, and N.D. Allen. 1993. Simple and efficient production of embryonic stem cell-embryo chimeras by coculture. Proc. Natl. Acad. Sci. USA. *90*:4582-4585.

64.Yang, J.T., T.A. Rando, W.A. Mohler, H. Rayburn, H.M. Blau, and R.O. Hynes. 1996. Genetic analysis of alpha 4 integrin functions in the development of mouse skeletal muscle. J. Cell Biol. *135*:829-835.

65.Zhou, Y., K.C. Lim, K. Onodera, S. Takahashi, J. Ohta, N. Minegishi, F.Y. Tsai, S.H. Orkin, M. Yamamoto, and J.D. Engel. 1998. Rescue of the embryonic lethal hematopoietic defect reveals a critical role for GATA-2 in urogenital development. EMBO J. *17*:6689-6700.

8 | Introduction of DNA into Plants

Michael A. Savka[1] and Andrew N. Binns[2]
[1]Department of Biological Sciences, Rochester Institute of Technology, Rochester, NY ; [2]Department of Biology, Plant Science Institute, University of Pennsylvania, Philadelphia, PA, USA

INTRODUCTION

The discovery that transgenic plants can be regularly generated via the utilization of *Agrobacterium tumefaciens* as the gene delivery system (6,126) marks the advent of plant genetic engineering. Since that time, an enormous amount of information on diverse approaches to introduce DNA into plant cells has been published. Additional approaches to introduce DNA into plant cells have been developed and include using polyethylene glycol (95), microinjection (29), electroporation (46,109), liposomes (44), plant viruses (15,58,62), and particle bombardment (77,78). The problems and strategies of introducing DNA into plant cells have recently been reviewed (13). The transfer and expression of modified plant and foreign genes into plants have revolutionized research in plant biology and have provided an avenue for the production of improved varieties of crop plants, novel ornamental varieties, and plant produced medicines (50,53,68,90).

The use of agrobacteria and particle bombardment have become the most common approaches to deliver DNA into plant cells. *Agrobacterium*-mediated gene delivery is the method of choice because it is amazingly more efficient than other methods (11), results in a higher frequency of single inserts than other methods (16,39,40,103), can theoretically deliver very large (>100 kbp) inserts (89), and is technically simple. Particle bombardment has been frequently used in cases, e.g., monocots, which are inefficiently transformed by *Agrobacterium*. The difficultly in employing *Agrobacterium* for monocot transformation is slowly being circumvented as information on the *Agrobac-*

Gene Transfer Methods
Edited by Pamela A. Norton and Laura F. Steel
©2000 Eaton Publishing, Natick, MA

terium T-DNA transfer mechanism accumulates (23,60,61). In this chapter we discuss laboratory protocols that illustrate *(i) Agrobacterium*-mediated, *(ii) in planta Agrobacterium*-mediated, and *(iii)* particle bombardment systems to introduce DNA into plant cells and regenerate transgenic plants.

GENE TRANSFER SYSTEMS

Classical *Agrobacterium* System

Over two decades ago, gene transfer between a soil bacterium *Agrobacterium* and plant cells was demonstrated (24). Since then, a wealth of information has been assembled on the *Agrobacterium* system as well as other means to introduce DNA into plant cells.

The *Agrobacterium*-mediated gene transfer system has proven to be the most successful system to create transgenic plants. This system is based on the natural capacity of the soil bacterium *A. tumefaciens* to transfer DNA (T-DNA) from a large tumor-inducing (Ti) plasmid into plant cells, resulting in the formation of tumors at the infection site. Various studies have shown that the T-DNA genes expressed in the plant cells encode proteins that catalyze the production of two classes of plant hormones, cytokinin, and auxin, which in turn are capable of stimulating plant cell division (12). The T-DNA also encodes a second set of genes involved in the synthesis of novel low molecular weight carbon and nitrogen compounds generically known as opines (36). The resultant tumor proliferates indefinitely on the plant and produces opines. The class of opine (e.g., octopine, nopaline, mannopine, agropine) synthesized in the transformed cells is dependent on the strain of *Agrobacterium* that initiated the tumor. Intriguingly, the strain of *Agrobacterium*, which induces the transformed cells, can also utilize the opines as a sole carbon and nitrogen source, thereby conferring on the *Agrobacterium* a selective advantage over other soil bacteria (92,104). Clearly, *Agrobacterium* can be considered a sophisticated parasite, capable of farming the plant that it transforms.

Agrobacterium-mediated transformation most efficiently transforms cells at a wound site on infected plants or plant tissues. Simultaneous with the attachment of bacteria to the plant cell wall, the *Agrobacterium* sense compounds released from wounded plant tissues. Specific active compounds, including some hexose sugars and phenolic compounds such as acetosyringone (19,111), induce the transcription of genes located in the virulence (*vir*) region of the Ti plasmid (125) that are required for the processing of the T-DNA from the Ti plasmid and transfer of the T-DNA into the plant cell. Briefly, VirD2 creates a single-stranded nick at 25-bp repeats (border repeats) that flank the T-DNA (122) and covalently binds to the 5′ end of the single-stranded DNA (ssDNA) molecule that is ultimately released from the Ti plasmid (Figure 1). T-strand formation is inititated at the right border (124) and proceeds from right to left. Thus, the T-DNA genes nearest to the right border are the first genes to be integrated into the nuclear genome of the plant cell (Figure 1). A key feature of plant genetic engineering is that any piece of DNA between the borders can be transferred to the plant cell. The actual transfer process is dependent on other Vir proteins, including the ssDNA binding protein, VirE2, and the 11 VirB proteins that are hypothesized to form a membrane-bound DNA transfer complex, the VirB pore (Figure 1). The reader is referred to reviews (5,26,28, 121,127) for a more detailed analysis of the transfer process.

Two additional technical advances have made *Agrobacterium* the standard method to generate transgenic plants. The first was the construction of plant

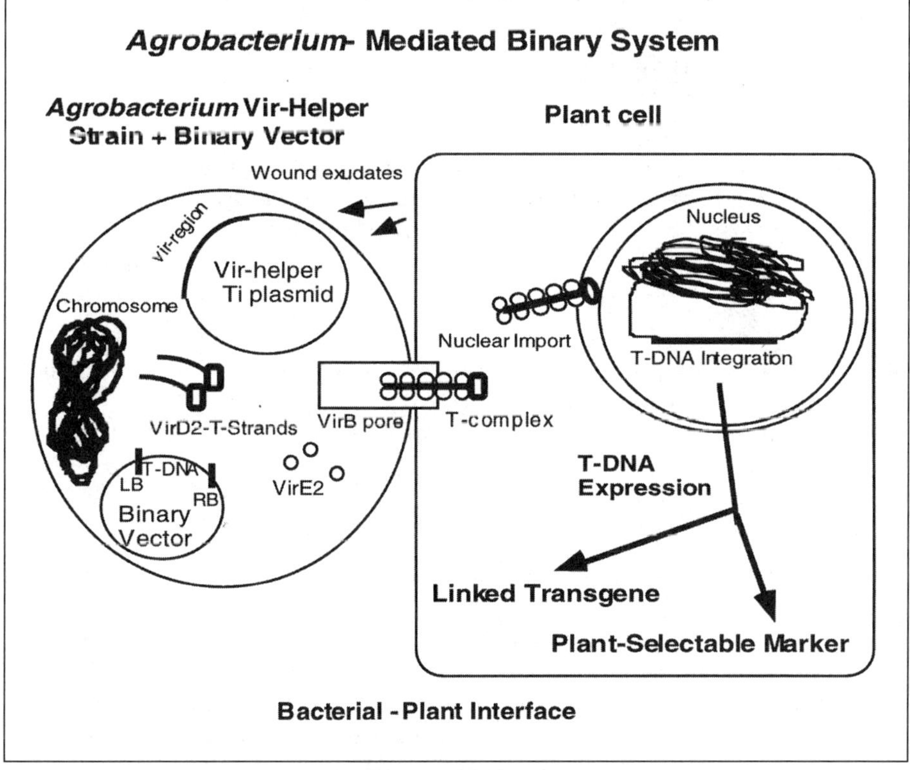

Figure 1. Schematic drawing of the basic steps in *Agrobacterium*-mediated delivery of transgenes into plant cells. Wound exudates, including acetosyringone (AS), activate the transcription of the virulence genes. VirD2 nicks at the right (RB) and left (LB) borders of the binary vector and covalently attaches to the 5′ nick site at the RB, forming the VirD2-T-strand. At some time during the transfer process, this single-strand DNA (VirD2-T-strand) intermediate interacts with the ssDNA-binding protein, VirE2, and is transported through the VirB pore into the plant cell. This T-complex moves into the plant nucleus where the T-DNA is randomly integrated into the chromosomal DNA and expressed.

expressable selectable markers (10,45). In this case, a plant active promoter is fused to a coding sequence that produces a protein capable of, for example, rendering the plant cell resistant to an antibiotic such as kanamycin. This type of selectable marker is required for all of the plant transformation protocols. The second crucial achievement necessary for routine *Agrobacterium*-mediated transformation was the demonstration (65) that a Ti plasmid carrying a deletion of the tumor causing T-DNA but retaining all of the virulence *vir* genes (the Ti helper plasmid) can function *in trans* to transfer DNA from a small plasmid containing border sequences flanking the DNA of interest (a binary vector). The combination of a selectable marker being moved into plants in the absence of genes (from the original T-DNA) that adversely affect the regeneration potential of the plant cell, and the ease of manipulating binary vectors so that they can be engineered to carry the genes of interest, allow the investigator to create transgenic plants rapidly and efficiently (Figure 2).

In Planta Agrobacterium System

The use of *Agrobacterium* infiltrated into seeds to produce transformed plants was originally developed in 1987 by Feldmann and Marks (38). Various modifications of this nontissue culture method of *Agrobacterium*-mediated plant transformation have been developed to introduce DNA into the emerging bolts of *Arabidopsis* (7,8,21,27). This method has become popular because it does not require expertise in plant tissue culture and is simple and reproducible with *Arabidopsis.* The *in planta Agrobacterium* system targets intact germ cells as recipients of the T-DNA, thus circumventing the possibility of somaclonal variation (81). The effect of somaclonal variation may mask the expression of the introduced gene of interest in the transgenic plant. Workers are investigating approaches to the *in planta Agrobacterium* system to transform important crop plants such as soybean.

Direct Gene Transfer by Particle Bombardment

The introduction of naked DNA into plant cells has been accomplished by particle acceleration into plant cells (77,78). This system is based on the acceleration of microprojectiles coated with DNA at forces capable of penetrating plant cell walls and membranes. The original biolistic gun used gun powder acceleration of tungsten or gold particles covered with DNA. The newer version of the particle gun uses helium acceleration of the DNA-coated particles. The different designs were compared by Johnston (72).

Particle bombardment has been used to introduce DNA into a wide variety of plant cell types and in many cases transgenic plants were recovered (Table 1). This approach has also proven successful for animal (116,123),

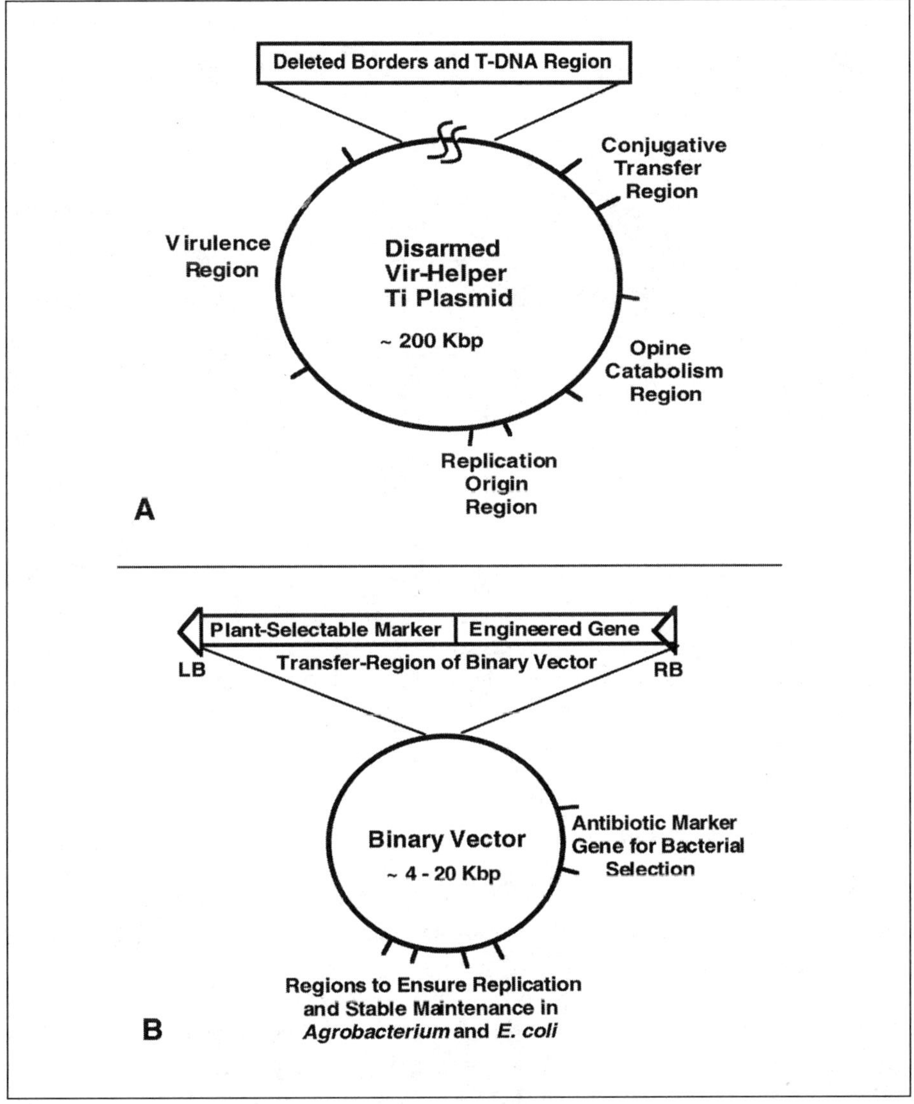

Figure 2. Schematic drawing of some features of the *Agrobacterium* binary vector system. (A) Disarmed Vir-helper Ti plasmid. The virulence region contains the *vir* genes required for T-DNA processing into a T-strand and assembly into the T-complex. Also shown are the deleted wild-type (auxin, cytokinin, and opine synthesis genes) border sequences and T-DNA region, the conjugal transfer region, the opine transport and catabolic region, and the origin of replication region. (B) Binary or T-DNA vector. The transfer-DNA region is the engineered T-DNA flanked by right border (RB) and left border (LB) sequences. The transfer region contains a plant-selectable marker gene to select transformed plant cells linked to an engineered gene of interest. The optimal configuration of the engineered transfer region is to have the gene of interest nearest to the RB and the selectable marker gene proximal to the LB. Since T-DNA processing and transfer begins at the RB and proceeds to the LB, this T-DNA orientation ensures that the selected plant cell will carry the linked gene of interest. Also shown are the antibiotic marker gene for bacterial selection of binary vector and the origin regions for *E. coli* and *Agrobacterium* replication.

Table 1. Some Plants Transformed by Microprojectile Bombardment

Plant	Explant Type	Tissues Produced	Reference
Tobacco	Cell Suspension	Plants	78
	Callus Cultures	Calli	102
	Leaves	Plants	77
Corn	Cell Suspension	Plants	47
	Embyonic Calli	Plants	47,52
Papaya	Immature Embryos	Plants	42
	Hypocotyl Segments	Plants	42
Norway Spruce	Embryogenic Calli	Calli	99
Poplar (hybrid)	Callus Cultures	Plants	85

green algae (14), and yeast (4) nuclear genomes and for plant chloroplast (17,31) and mitochondria (43,73) genomes. Furthermore, monocot (47) and gymnosperm (99) species that are recalcitrant to *Agrobacterium*-mediated DNA transfer have been successfully transformed by particle bombardment, and in the case of monocots, transgenic plants were recovered (Table 1).

PROTOCOLS

Classical *Agrobacterium* System

The classical *Agrobacterium*-mediated leaf explant transformation system was initially developed in 1985 by Horsch and coworkers (69) for *Solanaceae* species petunia and tobacco. Tobacco has developed as the model plant for studies of gene transfer using this gene transfer system. The advantages of *Arabidopsis thaliana* as the organism for plant molecular genetics has prompted the development of transformation protocols (83) using root (117), hypocotyledons (33), stem (48), and tubers explants (107). The classical *Agrobacterium* transformation system using leaf pieces has proven useful for petunia, tobacco, tomato, potato, and *Populus* sp. and relies on the totipotency of the targeted leaf cells. This system can be modified to accomodate other species if an efficient plant regeneration method is available for that species/cultivar. For example, explants other than leaves, such as petiole, stem, root, midrib, or cotyledons sections could be used in the classical *Agrobacterium* transformation system. The protocol described below is tailored for transformation of tobacco (*Nicotiana tabacum* cv.) leaf pieces using kanamycin or hygromycin as the plant cell-selectable marker.

Preliminary Considerations

Before beginning to produce transgenic plants via *Agrobacterium*, certain decisions must be made. These include the use of *vir*-helper strains, binary vector plasmids, plant selectable marker genes, and types of promoter to express the gene(s) of interest.

Vir-Helper Strains

Current approaches to use the classical *Agrobacterium* system usually employ a *vir*-helper strain to provide the virulence proteins *in trans* that are required for T-DNA processing, transfer, and integration into the nuclear genome of the plant cell (Figure 1). The most commonly used nononcogenic (disarmed) *vir*-helper plasmids derived from Ti plasmids are listed in Table 2. Disarmed *vir*-helper plasmids are derivatives of Ti plasmids in which the oncogenic T-regions have been deleted (Figure 2A). This eliminates the endogenous production of auxin or cytokinin directed by T-DNA genes in the transformed cell, which will interfere with transgenic plant regeneration. The helper strains listed in Table 2 have been derived from three Ti plasmid types and have been shown to be suited for a variety of plant species. For example, helper strain EHA105 has been proven useful in the transformation of legume species (66,67). No disarmed *vir*-helper plasmids have been derived from *A. rhizogenes* root-inducing (Ri) plasmids. Over the years, we have used all three helper strains listed in Table 2 to successfully transform tobacco leaf mesophyll cells and regenerate transgenic plants.

Wild-type Ri and Ti plasmids can be used to provide *vir* proteins *in trans* to binary vectors, but in some cases the transformed plant cells will contain the engineered T-DNA and the oncogenic T-DNA. This approach, however, has been shown to enhance the rate of transgenic shoot proliferation from leaf explants of a woody species, *Populus* (41). Likewise, wild-type Ri plasmids as the source of *vir* proteins have been useful in many cases to produce transgenic plants since the resulting hairy-root tissues often can be regenerated to fertile plants (57,96,97,114,115). Progeny of hairy-root regenerated plants can be screened for lines that contain only the engineered T-DNA (30,34).

Helper strains are used in combination with small plasmids (4–20 kbp) containing a broad range origin of replication that enables stable plasmid replication in either *Escherichia coli* or *A. tumefaciens*. These small plasmids, known as transfer-DNA (T-DNA) or binary vectors, contain 25 bp right and left border sequences between which a plant-selectable marker gene and other foreign genes of interest are cloned (Figure 2B). Right and left border sequences flanking the plant-selectable marker gene and other genes of interest in binary vectors provide the advantage that DNA insertions into the plant

Table 2. Disarmed *Vir*-Helper Strains Commonly Used in *Agrobacterium*-Mediated-Transformation Systems

Helper Strain	Helper Plasmid	Strain	Origin	
			Ti Plasmid	Reference
EHA105	pEHA105	A281 (succinamopine)	pTiBo542	66,67
GV3101	pMP90	C58 (nopaline)	pTiC58	79
LBA4404	pAL4404	Ach5 (octopine)	pTiAch5	65

genome will be limited to regions defined by these two border sequences (120,122). This feature is a major advantage of the *Agrobacterium*-mediated plant cell transformation system when compared to other approaches to introduce DNA into plant cells. In general, the binary vector is constructed and characterized in *E.coli* and then introduced into a *vir*-helper strain of *Agrobacterium*. The resulting helper strain containing a disarmed *vir*-helper plasmid and binary vector is also known as a binary *Agrobacterium* system to shuttle DNA into plant cells (Figure 1).

<u>Plant-Selectable and Plant-Screenable Marker Genes</u>

The T-DNA of the binary vectors must contain a genetic marker flanked by border sequences to enable transformed cells to be directly selected or screened. In most cases, workers have employed the antibiotics kanamycin and hygromycin as the antimetabolite to directly select for transformed plant cells containing the T-DNA encoded cognate detoxifying enzyme. Some commonly used plant-selectable and plant-screenable genes are listed in Table 3.

Binary vectors also contain multicloning sites that are 3′ of plant active promoters and 5′ of a polyadenylation signal for eukaryotic transcription initiation and termination, respectively. These multicloning sites facilitate the incorporation of a modified plant or foreign structural gene of interest. Over the last two decades numerous binary vectors for *Agrobacterium*-mediated transformation have been developed (2,9,32,55,63,64,70,71,76,84). Numerous binary vectors are now also available commercially (e.g., the pCAMBIA series from Cambia, Canberra, Australia).

<u>Chemically Inducible Promoters</u>

Conditional promoters have been developed more recently and are very useful for the study of gene regulation and the elucidation of gene function in var-

Table 3. Some Common Selectable and Screenable Genes Used to Recover and Identify Transformed Plant Cells

Plant-Selectable Genes			
Enzyme/Gene	Origin	Antimetabolite	Reference
Neomycin phospho-transferase II/*nptII*	*Tn 5*	Kanamycin	45
Phosphinothricin aceytltransferase/*bar*	*S. hygroscopicus*	Phosphinothricin	33
Hygromycin phospho-tranferase/*hyg*	*E. coli*	Hygromycin	118
Acetolactate synthase/*als*	Plants	Sulfonylureas	22
EPSP synthase/*aroA*	Plants and *E. coli*	Glyphosate	101,106
Dihydropteroate synthase/*sul*	pR46	Sulfonamide	56

Plant-Screenable Genes		
Enzyme/Gene	Origin	Reference
Green fluorescent protein/*fus*	*A. victoria*	20,59,94,105
β-Glucuronidase/*uidA*	*E. coli*	71,79
Luciferase/*lux*	Firefly	93
Luciferase/*luxA, B*	*V. harveyi*	80

ious plant biological processes (1,18,37,74,82,86,98,100). Some of the recently described conditional plant gene expression systems are listed in Table 4.

Preparation of Plant Material

The success of this approach relies on: *(i)* efficient transfer of genes from *Agrobacterium* to the plant cell, *(ii)* reliable selection for transformed plant cells, and *(iii)* regeneration of plants from selected transformed cells. An overview of the essential steps of this procedure is presented in Table 5 and detailed protocols are given below.

1a. Wild-type *Petunia hybrida* cv. or tobacco plants are grown from seed in a greenhouse or plant growth chamber. The plants are grown in a mixture of soil/vermiculite/peat moss (equal volume each). Environmental growth conditions should be about 26°C with fluorescent illumination for 16 to 18 hours per 24 hour period. Petunia plants approximately 15 cm in height bearing leaves about 5 cm in length give the best response. Vegetative tobacco plants approximately 25 to 30 cm in height bearing 7 to 10 expanded leaves yield the best results. Leaves (12–15 cm) that are medium green, thick, but not leathery, and with semi-opaque lower

Table 4. Some Chemically Inducible Plant Gene Expression Systems and Gene(s) Regulated in Transgenic Plants

Origin of Regulation	Chemical Inducer	Gene(s) Regulated	References
TetR repressor protein, *tet* operator sequences, CaMV 35S promoter sequences	Tetracyclines	*rolB* *abp1*	51,100 74
AlcR transcriptional activator, *alcA* binding sites, CaMV 35S promoter sequences	Alcohols	*gus2*	18
Yeast ACEI transcription factor, ACE1 binding domain, CaMV 35S promoter sequences	Copper	*ipt* *aatP2*	87,88 86
GAL4 binding domain, VP16 transacting domain, glucocorticoid receptor, GAL4 upstream activating sequence, CaMV 35S promoter sequences	Glucocortoids	*lux*	3

epidermis (white pubescence-like) will respond well in various tissue culture regimes.

1b. Alternatively, plants are grown under sterile conditions in tissue culture on plantlet medium (PM) in Magenta™ boxes (Sigma, St. Louis, MO, USA) (75–125 mL/vessel). Apical shoots from plantlets are transferred to fresh PM every 4 weeks for maintenance. Young and growing leaves from in vitro grown plantlets cultured under sterile conditions will respond well to plant transformation. In vitro grown plantlets do not require the surface-sterilization procedure given below before preparation and preincubation of leaf discs.

2. Leaves from plants greenhouse or plant chamber grown must be surface sterilized. Whole leaves are surface-sterilized for 10 minutes in a 0.525% aqueous solution of sodium hypochlorite (NaOCl) containing a drop of the surfactant Tween® 80 per 500 mL of disinfectant. The leaves are then rinsed two or three times with sterile distilled water for 5 to 10 minutes each wash.

3. The surface-sterilized leaves should be kept in a sterile petri plate (100 × 15 mm) containing about 7 mL of sterile water to prevent desiccation. All subsequent procedures are done using sterile technique in a

Table 5. Essentials of the Classical *Agrobacterium*-Mediated Transformation Protocol

Time	Procedure	Media	Important Considerations
Day 1	Cut leaf pieces.	SPM	Avoid dessication.
Day 2	Initiate overnight *Agrobacterium* culture.	LB plus antibiotics	Include appropriate antibiotic for binary plasmid selection.
Day 3	Select expanded leaf pieces.		Use expanded tissue for coculture with *Agrobacterium*.
	Pellet *Agrobacterium* and gently resuspend.		Be sure to remove antibiotic for binary plasmid selection.
	Submerge leaf pieces in resuspended *Agrobacterium*.		Cover explant with *Agrobacterium* and blot excess on sterile towels or filter papers.
	Place inoculated leaf discs in SPM without antibiotics.	SPM without antibiotics	Make sure leaf pieces are in contact with the medium.
Day 5	Wash infected leaf discs in LB with plant-selectable and bacteria-specific antibiotics.	LB with antibiotics	Blot washed discs to remove excess antibiotics.
	Transfer discs to SPM with plant-selectable and bacteria-specific antibiotics.	SPM with antibiotics	Make sure leaf pieces are in contact with the medium to ensure antibiotic selections.
Day 12; 7-day intervals	Transfer leaf discs to fresh SPM with plant-selectable and bacteria-specific antibiotics.	SPM with antibiotics	Transfer to promote rapid growth of transformed cells and to effectively control the *Agrobacterium* strain.
Antibiotic-resistant shoots present	Cut from leaf discs; place into RPM medium with plant-selectable and bacteria-specific antibiotics.	RPM with antibiotics	Transfer shoots 1 to 1.5 cm in length. Rooting should occur in 14 to 21 days. After roots present, transfer to PM with only bacteria-specific antibiotic.

horizontal laminar flow hood. Assure the sterility of the plant tissue in all subsequent steps.

Preparation and Preincubation of Leaf Discs

1. Leaf discs or pieces from surface-sterilized leaves or leaves of in vitro grown plants are cut with a 1.0-cm cork borer or scalpel. If hand cut with scalpel, the leaf pieces should be approximately 1.0 to 1.5 cm square. The center vein of the leaf is usually avoided because it is thick. Work quickly in the laminar hood to avoid desiccation of the leaf tissue. Desiccated leaf tissues will stick to each other and to scapel and forceps. The cut discs or pieces are placed abaxially (top surface of the leaf down) on the surface of shoot proliferation medium (SPM). It is important that the surface of the tissue is in direct contact with the medium. Allow condensation to evaporate by opening the lid slightly in the laminar flow hood. Condensation of water vapor in the Petri plate promotes contamination and vitrification, the latter leading to alter tissue growth (glasslike) and ultimately resulting in tissue loss.

2. After 48 hours of incubation on SPM at 26°C under fluorescent lighting (16–18-h photoperiod), about 50% of the leaf discs or pieces should have enlarged to about twice their original size. The leaf discs that expand are the tissues that will respond most favorably to *Agrobacterium* transformation. The expanded leaf disc explants are selected for coculture with *Agrobacterium* strains.

Preparation of Agrobacterium *Cultures*

1. For tissue inoculation, grow up separate overnight bacterial cultures (5–10 mL) in Luria broth (LB) of the *Agrobacterium* strains EHA105 (the control) and EHA105 containing a binary vector construct (the experimental).

2. It is necessary to remove antibiotics after overnight growth of the bacterial strains. For example, strain EHA105 containing a binary vector is grown in the presence of an antibiotic for the maintenance of the binary vector. If the antibiotic kanamycin (Km) is used, Km must be removed from the bacterial culture before inoculating plant tissues because it is toxic to plant cells. To remove the antibiotic from the culture, pellet bacteria in a microcentrifuge by spinning at 8000 rpm for 12 minutes, aspirate the spent broth, and resuspend the bacterial cells in fresh liquid SPM, Murashige and Skoog medium (MS) or LB without antibiotics. The bacterial pellets are resuspended in equal volume of LB by agitation

with sterile toothpicks or small wooden dowels; avoid vortex mixing the pellet to resupsend. All of these procedures should be performed using sterile materials and technique in a horizontal laminar flow hood.

3. Transfer each resuspended culture to a sterile Petri dish. The bacterial density of the overnight culture is reduced by one half. For example, if the resuspended pellet is from a 5-mL overnight culture, add 5 mL of additional LB to the resuspended cells. A very concentrated suspension of *Agrobacterium* as the inoculation medium can severely damage the explants during the coculture period because of rapid overgrowth on the explants. The expanded preincubated leaf explants are mixed, to randomize differences among individual plants and the origin of leaf position, and are submerged in the overnight culture for the time required to transfer them to fresh SPM plates. Be sure to allow the liquid culture to completely cover the entire leaf piece edge, as it is the cells in the wounded edge and adjacent mesophyll cells that are the potential target cells for transformation events.

4. Blot the infected pieces individually on sterile paper towels or filter paper to remove excess resuspension medium. Transfer the infected leaf explant abaxially onto the surface of a fresh SPM plate. Make sure leaf discs are in good contact with the surface of the medium by pushing the tissue down into the medium with forceps. In certain cases, the addition of a *vir* gene inducer, acetosyringone (AS) (111,112), at 10 to 100 μg/mL to the SPM medium for coculture (the culture of plant leaf explants with binary *Agrobacterium* strain in the absence of selection) will enhance transformation frequencies from 4- to 8-fold with tobacco leaf pieces (11,108). In addition, if working with a plant species not previously transformed by *Agrobacterium*, the addition of AS to the coculture medium should enhance the transformation process.

Transfer of Leaf Discs to Selection Plates

1. Following 36 to 48 hours of coculture, the *Agrobacterium* must be prevented from overgrowing the explants. The leaf disc explants (usually 12) containing the same *Agrobacterium* strain are washed in liquid (15 mL) SPM, MS, or LB medium in a sterile beaker containing a plant-selectable marker such as Km at 50 to 100 μg/mL, timentin (Tm; 200 μg/mL, ticarcillin disodium and clavulanate potassium) or another antibiotic to suppress *Agrobacterium*, such as cefotaxium (Cf; 400 μg/mL) or carbenicillin (Cb; 200 μg/mL). A combination of two bacteria-specific antibiotics is a very effective approach to control the *Agrobacteium* binary strain. A study of the effects of combining various antibiotics on

Agrobacterium and tobacco cell suspension is a useful reference (25). The leaf discs in this medium are rapidly swirled with sterile forceps to remove excess *Agrobacterium* that has grown on the leaf pieces. In our hands, Tm at 200 to 300 µg/mL is most effective in the suppression of *Agrobacterium* in comparison to Cf or Cb.

2. The washed discs are blotted on sterile paper towels or filter paper and transferred to SPM supplemented with Tm (200 µg/mL), Cf (400 µg/mL), or Cb (200 µg/mL) to suppress the growth of the T-DNA donor *Agrobacterium*. Also add to SPM the appropriate plant-selectable antibiotic (for T-DNA which introduces the *nptII* gene into transformed plant cells, use Km at 100–125 µg/mL) to select for stable transformed plant cells. If the T-DNA contains the *hptII* gene, use hygromycin B (Hyg) at 12 µg/mL as the plant-selectable antibiotic. It is important to position the tissue pieces or discs into the regeneration/selection medium well enough to allow direct contact, which maximizes antibiotic selection for transformed plant cells and for the suppression of *Agrobacterium*.

3. Leaf discs are transferred to SPM with fresh antibiotics every week. If leaf discs become covered with persistent agrobacterial growth, the discs can be washed in the presence of antibiotics a second time (Step 1, above). Blot and transfer washed leaf pieces to fresh SPM with antibiotics for *Agrobacterium* and transformed plant cell selection.

Evidence of Transformed Plant Cells

1. After approximately 2 to 4 weeks, regions (knobs) or zones of green masses of cells (callus tissue) will be evident on the edge of the leaf explants. The callus regions will appear as a confluent or semiconfluent mass of callus tissue on the leaf tissue edge or as individual callus colonies along the edge. The growth of callus tissue on the edge of the leaf pieces indicates recovery of transformed plant cells.

2. Leaf discs that contain Km- or Hyg-resistant callus are transferred to fresh SPM with Tm or Cf or Cb and the plant-selectable antibiotic at 14-day intervals or as needed to suppress *Agrobacterium* growth. It is important not to allow condensation to accumulate in the culture dishes. Free standing water will favor vitrification and thus loss of transformed tissues.

Plant Regeneration

1. After 2 to 3 additional weeks, shoot meristems will differentiate from antibiotic-resistant calli. This is evident by the formation of a green node and subsequent growth of a shoot meristem from the node (Figure 3).

2. Carefully cut regenerated shoots that are 1 to 1.5 cm or larger from the callus tissue, and place the basal cut stem into root proliferation medium (RPM) containing Km (50 µg/mL) or Hyg (6 µg/mL) and, for example, Tm to continue to suppress *Agrobacterium*. RPM is PM with the antibiotic Km or Hyg and dispensed into petri plates (25 mL/plate). The absence of plant growth regulators in PM favors root differentiation from the cut stem, and the presence of the plant-selectable antibiotic allows for a second selection of transgenic shoots.

3. Putative transgenic shoots that develop a root in the presence of the plant-selectable antibiotic (secondary selection) will be scored as antibiotic-resistant transgenic plants. If the regenerated shoots are not chimeric (composed of a combination of transformed and wild-type cells), rapid root development will be evident within 14 to 21 days on RPM (Figure 4). After roots are evident, carefully transfer transgenic plantlets to PM containing an *Agrobacterium*-specific antibiotic such as Tm. One transgenic plant per Magenta vessel enables rapid growth.

Confirmation of Transformation

Before a detailed analysis of transgene expression is carried out, it is important to establish the number of T-DNA insertions and perhaps the extent of rearrangements within the T-DNA in the antibiotic-resistant transgenic plants. Briefly, this can be performed by genomic Southern hybridization analyses (110) of introduced DNA with probes from different regions of the T-DNA.

1. Total DNA of the transgenic plants is isolated and purified by commonly used methods such as cetyltrimethylammonium bromide (CTAB)-based protocol (113) or by using currently available commercial kits. The total DNA is digested with a restriction endonuclease that will cut the T-DNA at a single internal site. The resulting restricted DNA is then fractionated by agarose gel electrophoresis and transferred to a membrane. DNA from nontransgenic wild-type plants is also isolated, restricted with the same restriction endonuclease, fractionated, and transferred.

2. A DNA hybridization probe derived from the T-DNA vector including the right border (RB) sequence and a few hundred base pairs of sequence flanking inside the T-DNA will estimate T-DNA copy number in the genome of the transgenic plant. If a single band hybridizes with this RB probe, this suggests a single integrated T-DNA in the transgenic plant. Additional analyses may include the use of internal T-DNA fragments or simultaneously two hybridization probes, for the right border and left border region, to more clearly define the limits of the integrated T-DNA.

During the introduction of T-DNA into the nuclear chromosome, structural rearrangements may occur such as deletions and/or the formation of inverted or tandom repeats of the T-DNA (35,75). The T-DNAs can be genetically linked or located on separate chromosomes. The results of the deletions and multiple insertions may lead to significant quantitative variation of transgene expression among independent transformants. Further effects on the level of transgene expression, known as position effects, are due to plant sequence context and/or higher order genetic structure flanking the T-DNA insertion. It is most desirable to work with transgenic plants containing a single full-length T-DNA insertion. The presence of multiple copies of homologous transgenes can lead to suppression of some or all of the gene copies. This phenomenon is termed gene silencing, and although the mechanisms involved have not been clearly defined, gene silencing can be the underlying factor in transgenic plants that exhibit unpredicted transgene expression (54,119).

Figure 3. Regeneration of transgenic tobacco shoots from leaf pieces using the classical *Agrobacterium* transformation system. Formation of nodes and growth of transgenic shoots from nodes in the presence of the antibiotic on leaf pieces inoculated with a Vir-helper strain carrying a binary vector conferring resistance to hygromycin (left and bottom). Leaf pieces (right and top) inoculated with Vir-helper strain only.

174

Additional analyses of transgenic plants may include the segregation of the antibiotic resistance selectable gene in selfed progeny seedlings. This is accomplished by germinating surface sterilized R1 seed on PM containing the plant-selectable marker antimetabolite. After 3 to 5 weeks, the seedlings that appear bleached or white are the antibiotic-sensitive progeny, whereas the green and actively growing seedlings inherited the plant-selectable marker gene and are thus antibiotic resistant (Figure 5). Statisitical analysis using the chi-square test can predict the number of independent segregating loci (38).

After a few independent transformants are chosen for further work, standard RNA (Northern) and protein (Western) blots can be performed as well as enzyme assays to further characterize the expression of the transgene. Semiquantitative reverse transcription polymerase chain reaction (RT-PCR) can be used to characterize messenger RNA levels and to identify individual transformants for additional investigations.

In Planta Agrobacterium System for *Arabidopsis*

A summary of the essential steps for the *in planta Agrobacterium* transformation of *Arabidopsis* is found in Table 6. A detailed protocol is given below.

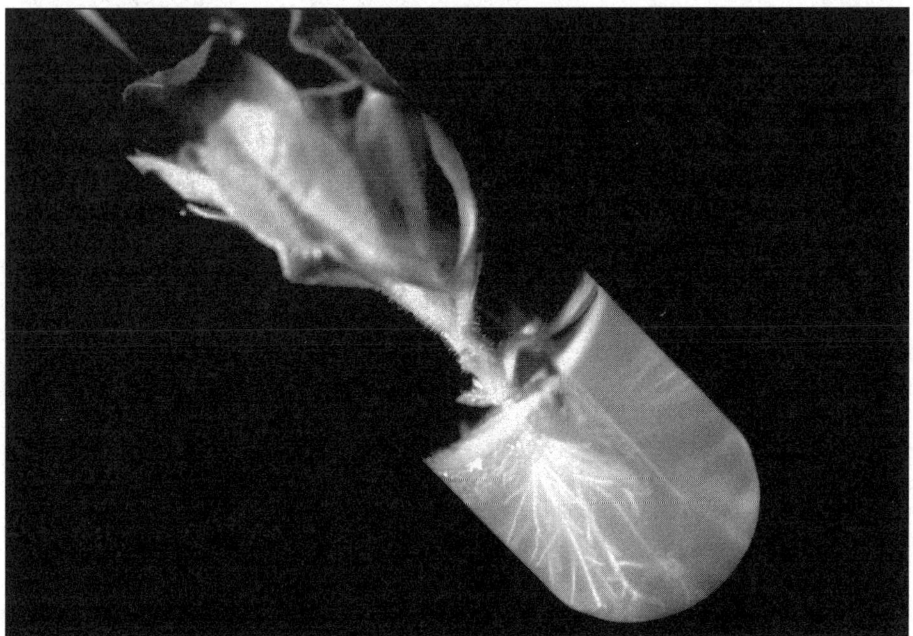

Figure 4. Root formation on an antibiotic-resistant shoot. Rapid root development on the basal region of a hygromycin-resistance shoot in the present of the antibiotic (second selection). This plantlet has successfully regenerated a shoot and root system in the presence of hygromycin and is ready to be transferred to a Magenta or similar vessel containing PM medium with Tm.

Plant Growth

1. Grow *Arabidopsis* plants of the ecotypes Columbia and Landsberg erecta to a stage at which bolts are just emerging. In general, about a dozen plants are grown in a 3.5-inch pot containing a commercial potting mixture. The pots are covered with a nylon screen or cheesecloth to retain soil when inverted for infiltration. To achieve a healthy plant that produces a good seed yield, it is best to grow plants under a short day period (9 hours light/day at 22°C) for the first 3 to 4 weeks. Plants can be grown for an additional 2 weeks to delay the date of the infiltration treatment. Then transfer plants to long days (16 hours light/day at 22°C) to induce flower bolting.

2. Once the primary inflorescence shoots have reached about 8 cm, clip off emerging bolts to encourage growth of multiple secondary bolts. Vacuum infiltration should be done 4 to 8 days after clipping the bolts. The physiological condition of the plants is very important. Plants with leafy

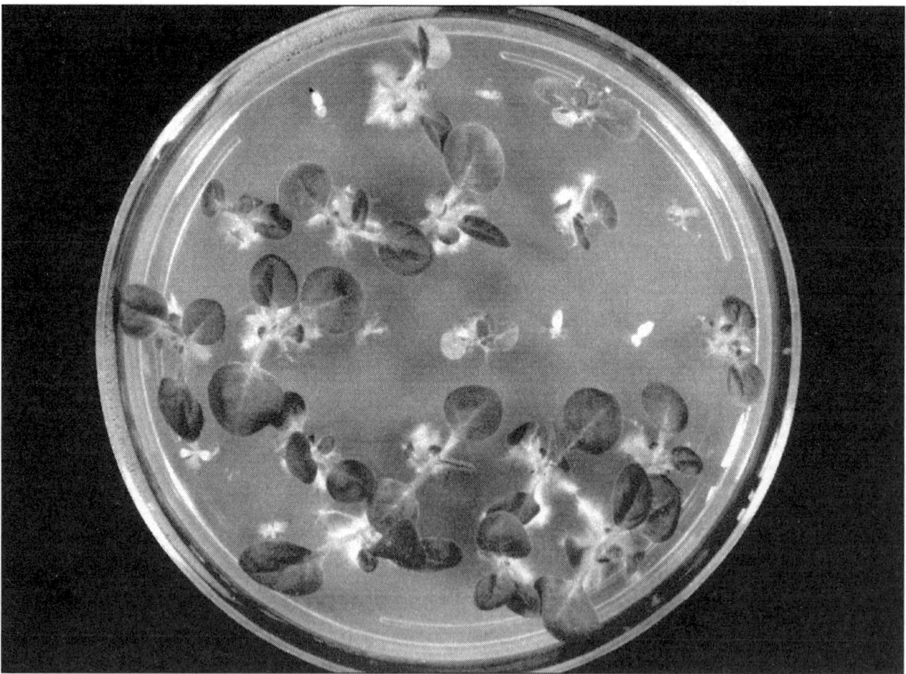

Figure 5. Segregation of a plant-selectable marker gene in progeny plants. First generation selfed seed from a transgenic plant were germinated on PM medium containing the plant-selectable marker antimetabolite. Seedlings that inherited the plant-selectable marker germinated and grew to green plants with well-developed leaves and root system on PM medium containing the antibiotic. Seedlings that appear bleached or white are the progeny that are sensitive to the antibiotic and failed to inherit the antibiotic resistant marker gene.

Table 6. Essentials of the *In Planta Agrobacterium*-Mediated Transformation Protocol

Time	Procedure	Media	Important Considerations
Day 1	Transfer plants to 16 hours light/day at 22°C		Plants should be grown from seed for 3 to 6 weeks in short days (9 hours of light/day).
Day 6	Cut off primary inflorescence (bolts).		Cutting off primary inflorescence will favor the growth of multiple secondary bolts.
Day 10–14	Prepare overnight culture of the *Agrobacterium* helper strain carrying the appropriate gene construct in a binary vector.	LB with antibiotics	Grow at 28°C with shaking overnight.
Day 11–15	Inoculate a 500-mL culture with overnight culture.	LB with antibiotics	Grow at 28°C with shaking overnight.
Day 12–16	Harvest *Agrobacterium* cells and resuspend in infiltration media.	5% sucrose solution	Resuspend *Agrobacterium* in infiltration medium to an OD_{600} of 0.8.
	Submerge plants in infiltration medium containing *Agrobacterium* and apply a vacuum to 16 mbars or until bubbles form, then release vacuum.		An alternative to the infiltration step is to use the surfactant Silwet L-77.
	Grow plants for about 4 more weeks.		Grow plants under long day periods (16 hours of light/day at 22°C)
After siliques are beginning to dry	Surface sterile seed and air-dry.		70% ethanol and 50% bleach solution with Tween 80 and rinse with distilled sterile water.
	Plate seed on selection plates and place at 4°C for 2 to 3 days.		Best to plate about 1000 seeds per 15×100-mm plate.
	Place plates at 22°C under long day light (16 hours light).		
	After 10 to 14 days, antibiotic-resistant transformants will be identifiable as dark green plants with growing roots.		

green rosettes respond best to transformation. This is especially true for ecotype Landsberg erecta.

Vacuum Infiltration and Seed Production

1. Prepare a 25-mL overnight liquid culture with appropriate antibiotics of *Agrobacterium* carrying the appropriate gene construct(s) in a binary vector. The next day, inoculate a 500-mL liquid culture with antibiotics with the 25 mL overnight culture.

2. The next day, harvest cells by spinning in a high-speed centrifuge at 8000 rpm for 12 minutes at 4°C. Resuspend in 3 volumes of infiltration medium (5% sucrose solution) to an optical density (OD_{600}) of approximately 0.8. Harvested cells at an OD_{600} of 2.0 will yield enough *Agrobacterium* cells for infiltration of 6 to 8 pots. The OD can be determined with a spectrophotometer.

3. Add *Agrobacterium* (in infiltration medium) to a dish or beaker and invert potted plants into liquid solution. It is important to submerge flower bolts and entire rosettes. A 1 L beaker filled with greater than 200 mL of solution fits well with 3.5-inch pots. The bacterial infiltration solution can be used to inoculate at least one additional potted plant. The frequency of transformation does not decrease between the first and second pot inoculated.

4. Place the inoculated potted plant into a bell jar. Draw a vacuum to about 16 mbars or until bubbles form on leaf and stem surface and solution starts to bubble a bit (approximately 5 min), then release the vaccum rapidly. Sufficient infiltration is visible as uniformly darkened water-soaked tissue. Be sure to have good vacuum traps in the system to avoid contaminating the vacuum pump oil with medium. After the infiltration treatment, the plants will appear glassy and dark green in color.

   An alternative to vacuum infiltration in this approach is to use a surfactant. Before inverting the pot into the *Agrobacterium* medium, add Silwet L-77 (0.05%). The surfactant Silwet L-77 (Lehle Seeds, Round Rock, TX, USA) is toxic at times, and a lower concentration of the surfactant may be necessary (0.02% or 0.005%). Dip the above-ground portions of the plant in the *Agrobacterium* solution containing Silwet L-77 for 2 to 3 seconds with gentle agitation. It is best to completely dip the above-ground portion to inoculate the lower axillary inflorescences.

5. After vacuum infiltration or treatment with the surfactant, lay potted plants on side and cover with plastic wrap to maintain high humidity. Keep plants under high humidity conditions for 24 to 48 hours, then remove the plastic wrap, and place upright.

6. Grow plants under long day periods (16 hours light/day at 22°C) for appoximately 4 weeks and keep the bolts from each pot separate. When the siliques on the plants begin to dry, or the basal siliques start to open, harvest all the seeds from one pot together.

Selection of Antibiotic-Resistant Transformants

1. The seeds are surface-sterilized in a 15-mL plastic tube by treating them with 70% ethanol for 2 minutes, soaking for 5 minutes in 50% bleach solution with Tween 80 (0.01%), followed by 3 rinses in sterile water. Seeds in the tube can be air-dried in a horizontal laminar flow hood before plating on selection medium.

2. Prepare selection medium with the plant-selectable antibiotic (e.g., Km at 50–100 µg/mL) and bacterial-specific antibiotics such as Tm at 200 µg/mL to suppress *Agrobacterium* growth. It is customary to add 25 mL of medium into a 15 × 100-mm Petri plate.

3. Plate seeds by resuspending in sterile water and spreading onto the selection medium. Plates containing seeds are then dried in a laminar flow hood to evaporate free standing water. Also, place a few seeds from an authentic transformed plant containing the same plant-selectable marker as the positive control. It is best to plate about 1000 seeds onto selection medium in a 15 × 100-mm plate. Higher seed density will lead to false positives or escapees, which appear to be antibiotic resistant.

4. After plating seeds, cold treat the plated seeds for 48 to 72 hours in a cold room or refrigerator and then place in growth chamber under long day periods of light (16 hours) at 22°C.

5. After about 10 to 14 days, antibiotic-resistant transformants will be identifable as dark green plants with fully developed cotyledons and well-developed roots that have grown into the selection medium. The frequency of transformation is on average between 0.2% and 0.8%.

6. Healthy plants are transferred to soil, grown to maturity, and seed collected. Agar from the plant root system is removed by gentle washing with tap water. The soil should be saturated with water before plant transfer, and the environment around the leaf tissue kept at a high humidity by using plastic wrap or a dome. Allow plants to harden-off for about 5 days with high humidity in a growth chamber. Then remove the plastic wrap or dome and allow plants to grow to maturity.

Confirmation of Transformation

Follow procedure for the classical *Agrobacterium*-mediated transformation system previously described. Segregation data suggests that about 50% of the first generation plants contain a single T-DNA locus. The number and organization of the multiple T-DNA copies in single T-DNA locus is not known, but can be investigated as described in the procedure for the classical *Agrobacterium* transformation system.

Particle Bombardment

Preparation of Apparatus and Microparticles

This protocol is for helium acceleration of gold particles coated with DNA. Please interface the following protocol with manufacturer's recommendations and safety procedures.

1. Sterilize accelerator chamber and tube with 70% ethanol.
2. Sterilize rupture (1100 psi) and flying disks by soaking in 70% ethanol for 5 minutes. Air-dry in laminar flow hood.
3. Sterilize macroprojectiles and stopping plates by soaking in 70% ethanol for 5 minutes followed by air-drying in a horizontal laminar flow hood.
4. The DNA carrying particles can either be tungsten or gold microparticles between 1 and 10 μm in diameter. Gold particles are usually about 1 μm in diameter and must be used when using helium as the propelling gas.
5. To precipitate DNA onto gold particles, add 20 μg of CsCl-purified DNA (see Chapter 2 in this book) to 20 μL of TE buffer (10 mM Tris, 1 mM Na$_2$EDTA, pH 8.0). This will give a final concentration of 1 μg/μL.
6. Combine 100 mg of particles with 1.0 mL of TE buffer.
7. Add together 100 μL of particles from step 6 with 20 μg of DNA from step 5, 100 μL of CaCl$_2$·2H$_2$O of a 2.5 M stock, and 40 μL spermidine ·3H$_2$O of a 0.1 M stock.
8. Incubate at -70°C for 5 minutes to assist DNA precipitation, then spin in a microcentrifuge to pellet particles and DNA for 15 minutes at 10 000 rpm.
9. Decant supernatant with micropipettor. The remaining slurry of particles and DNA is to be used for bombardment. Two to three microliters of the slurry of particles is used per bombardment shot.

Introduction of DNA into Plant Cells

1. Clean chamber of the apparatus with 70% ethanol.
2. Spread the plant tissue over a 3-cm square area in a standard Petri dish.

The age and type of tissue will greatly affect the results, and optimization must be done for each species and tissue type. The tissue can be bombarded over filter paper or on tissue culture medium. If leaf explants will be bombarded, a screen is used to immobilize the tissue during bombardment.

3. Place the Petri dish in the chamber of the microprojectile accelerator about 14 cm from the macroprojectile stopping plate. Make sure the stopping plate is in the holder and the retaining screws are tightened. Use caution.

4. Vortex mix 2 to 3 μL slurry of DNA-particle mixture from step 9 (above). Resuspend the particles in 30 μL of absolute ethanol. Pipet 5.0 μL onto the macroprojectile or flying disk. The disk containing the slurry is then placed in the holder and air-dried for 5 minutes. The samples may be stored in a box above the silica dessicant Drierite™ (anhydrous $CaSO_4$; Mallinckrodt Baker, Phillipsburg, NJ, USA). Storage above Drierite is important to obtain reproducible results if air humidity is higher than 50%.

5. Turn helium tank on. Set delivery pressure in regulator for 1100 to 1300 psi.

6. Turn on vacuum pump and place rupture disk in retaining cap and screw in place.

7. Put flying disk in place with face down and place assembly in chamber.

8. Place plant tissue sample into chamber and close door.

9. Open vacuum valve. When vacuum reaches 25 to 28 inches of Hg, press fire button. Keep pressing until a noise is heard, which indicates the rupture disk is being broken. Use caution.

10. Release vacuum and remove sample.

11. Start with step 6 to repeat the operation on additional tissue samples.

This approach to introduce DNA into plant cells has two sets of variables that need to be optimized. Physical variables include the size and type of particles, the distance of the stopping plate, the distance of the sample from the stopping plate, the strength of the vacuum, and the disposition of the tissue on the Petri dish. Biological variables include the vector construct that dictates the antimetabolite to select for transformed plant cells. Other variables include the quantity of DNA, the cell cycle of the plant tissues, and the subculture cycle of the tissue samples.

Confirmation of Transformation

If stable transformants are desirable, antibiotic selection is initiated 48 hours after microprojectile bombardment. Depending upon the presence of screenable genes in the binary vector, various approaches can be considered, such as transient assays. To confirm that the introduced DNA is in the genome of the plant cells, refer to the discussion on confirmation of transformation in the classical *Agrobacterium* transformation system.

MEDIA, SOLUTIONS, AND REAGENTS

Classical *Agrobacterium*-Mediated Transformation

Shoot Proliferation Medium

For 1.0 L of medium:

700 mL	Distilled water
100 mL	MS macro nutrients stock (10×)
10 mL	MS micro nutrients stock (100×)
5 mL	MS iron stock (200×)
10 mL	MS vitamin stock (100×)
30 g	Sucrose

Adjust pH to 5.7 to 5.8. Adjust volume with distilled water to 1.0 L. Add 6 to 8 g agar.

Autoclave, cool to approximately 45° to 50°C then add:

Plant hormones: 1.0 mL of kinetin or BAP stock and 0.1 mL of NAA stock (see below and section entitled Plant Hormone Stock Solutions for components and concentrations for these solutions).

Antibiotic for plant-selectable marker (e.g., Km sulfate at 50 μg/mL).

Dispense: 25 mL per 15 × 100-cm Petri plate, cool in laminar flow hood.

An auxin α-naphthalene acetic acid (NAA; 0.1 mg/L) and a cytokinin kinetin (K; 1.0 mg/L) are supplemented to induce plant cell division and shoot organogenesis. Dispense 25 mL of medium per Petri plate (100 × 15 mm) containing 25 μL of K and 2.5 μL of NAA and allow to cool with lid slightly open in the laminar flow hood to avoid condensation. Condensation of water vapor in the Petri plate promotes contamination and vitrification, the latter leading to alter tissue growth (glasslike) resulting in tissue loss.

Root Proliferation Medium

For 1.0 L of medium:

700 mL	Distilled water
100 mL	MS macro nutrients stock (10×)
10 mL	MS micro nutrients stock (100×)
5 mL	MS iron stock (200×)
10 mL	MS vitamin stock (100×)
10 g	Sucrose

Adjust pH to 5.7 to 5.8. Adjust volume with distilled water to 1.0 L. Add 6 g agar.

Autoclave, cool to approximately 45° to 50°C then add:

Antibiotic for plant-selectable marker (e.g., Km sulfate at 50 µg/mL).

Antibiotic for *Agrobacterium* counter selection (suppression) (e.g., Tm at 200 µg/mL).

Dispense: 25 mL per 15 × 100-cm Petri plate, cool in laminar flow hood.

Plantlet Medium for Magenta Boxes

For 1.0 L of medium:

800 mL	Distilled water
50 mL	MS macro nutrients stock (10×)
5 mL	MS micro nutrients stock (100×)
2.5 mL	MS iron stock (200×)
10 g	Sucrose

Adjust pH to 5.7 to 5.8. Adjust volume with distilled water to 1.0 L. Add 6 g agar.

Autoclave, cool to approximately 45° to 50°C, then add:

Antibiotic for *Agrobacterium* selection (suppression) (e.g., Tm at 200 µg/mL).

Dispense 75 to 125 mL of PM per Magenta GA-7 vessel and cool.

MS (91) Macro Nutrients: 10× Stock Solution

For 1.0 L:

NH_4NO_3	16.5 g
KNO_3	19.0 g
$MgSO_4 \cdot 7H_2O$	3.7 g
$CaCl_2 \cdot 2H_2O$	4.4 g
KH_2PO_4 (monobasic)	1.7 g

Stock solutions of salts and iron are sterilized by autoclaving and stored at 4°C.

MS Micro nutrients: 100✕ Stock Solution

For 1.0 L:

H_3BO_3	0.6200 g
$MnSO_4 \cdot 7H_2O$	1.5640 g
$ZnSO_4 \cdot 7H_2O$	0.8600 g
KI	0.0830 g
$Na_2MoO_4 \cdot 2H_2O$	0.0250 g
$CuSO_4 \cdot 5H_2O$	0.0025 g
$CoCl_2 \cdot 6H_2O$	0.0025 g

Iron: 200✕ Stock Solution

For l.0 L:

$Na_2EDTA \cdot 2H_2O$	37.24 g
$FeSO_4 \cdot 7H_2O$	27.80 g

Connect tubing through cotton-filled trap to air line. Cover with foil and bubble and stir vigorously for about 5 hours or until completely dissolved. A clear dark brown solution will be produced.

MS Vitamins: 100✕ Stock Solution

For 1.0 L:

Thiamine·HCL	0.01 g
Pyridoxine·HCL	0.05 g
Nicotinic acid	0.05 g
Inositol	10.00 g

Vitamin stock solutions are filter-sterilized and stored at -4°C.

Plant Hormone Stock Solutions

<u>Auxin Stock Solution: 1.0 mg/mL</u>

For 100 mL:

One hundred milligrams of NAA is dissolved in 500 μL of 95% ethanol or 0.1 *N* KOH. Then slowly add distilled water to 100 mL.

<u>Cytokinin Stock Solution: 1.0 mg/mL</u>

For 100 mL:

> One hundred milligrams of K or 6-benzylaminopurine (BAP) are dissolved in 500 µL of 0.5 N HCl. Then slowly add distilled water to 100 mL.

> Sterilize auxin and cytokinin stock solutions using a 0.2-µm membrane filter and store at -4°C or colder.

Antibiotics

Kanamycin A monosulfate (Km; stock of 100 mg/mL; Sigma)

Hygromycin B (Hyg; stock of 50 mg/mL, Sigma)

Cefotaxime, sodium salt (Cf; stock of 200–300 mg/mL, Life Technologies, Rockville, MD, USA)

Carbenicillin, disodium salt (Cb; stock of 200–300 mg/mL, Sigma)

Timentin (Tm; stock of 200–300 mg/mL; ticarcillin disodium and clavulanate potassium; 30:1; SmithKline Beecham Pharmaceuticals, Philadelphia, PA, USA)

All of these antibiotics are water soluble. Prepare the desired stock concentration, sterilize using a 0.2-µm membrane filter, and store at -4°C or colder.

In Planta Agrobacterium-Mediated Transformation

Infiltration Medium

For 1.0 L of medium:

800 mL	Distilled water
50 mL	MS macro nutrients stock (10×)
5 mL	MS micro nutrients stock (100×)
2.5 mL	MS iron stock (200×)
10 mL	B5 vitamin stock (100×)
50 g	Sucrose

Adjust pH to 5.5. Adjust volume with distilled water to 1.0 L.

Add 6 to 8 g agar.

Autoclave, cool to approximately 45° to 50°C, then add:

> Plant hormones: Cytokinin, BAP at a final concentration of 0.044 µm (10 µL/L of a 1 mg/mL stock, see section entitled Plant Hormone Stock Solutions).

Dispense: 25 mL per 15 × 100-cm Petri plate, cool in laminar flow hood.

Selection Plates

For 1.0 L of medium:

800 mL	Distilled H_2O
50 mL	MS macro nutrients stock (10×)
5 mL	MS micro nutrients stock (100×)
2.5 mL	MS iron stock (200×)
10 mL	B5 vitamin stock (100×)
10 g	Sucrose

Adjust pH to 5.5. Adjust volume with distilled water to 1.0 L.

Add 6 to 8 g agar.

Autoclave, cool to approximately 45° to 50°C then add:

Antibiotic for plant-selectable marker (e.g., Km sulfate at 50 μg/mL) and antibiotic Cf or Tm to suppress *Agrobacterium* growth.

Dispense: 25 mL per 15 × 100-cm Petri plate, cool in laminar flow hood.

B5 Vitamins (49): 100× Stock Solution

For 1.0 L:

Thiamine·HCL	1.0 g
Pyridoxine·HCL	0.1 g
Nicotinic acid	0.1 g
Inositol	10.0 g

ACKNOWLEDGMENTS

Support for this work was provided by the National Science Foundation (Grant No. MCB-9817149) and the National Institutes of Health (Grant No. GM47369) to A.N.B. and by the National Science Foundation's Division of Undergraduate Education (Grant No. DUE9850657) to M.A.S.

REFERENCES

1. Ainley, M.W., K.J. McNeil, J.W. Hill, W.L. Lingle, R.B. Simpson, M.L. Brenner, R.T. Nagaoand, and J.L. Key. 1993. Regulatable endogenous production of cytokinins ups to "toxic" levels in transgenic plants and plant tissues. Plant Mol. Biol. *22*:13-23.

2. An, G. 1986. Development of plant promoter expression vectors and their use for analysis of differential activity of nopaline synthetase promoter in transformed tobacco tissue. Plant Physiol. *81*:86-91.

3. Aoyama, T. and N.-H. Chua. 1997. A glucocorticoid-mediated transcriptional induction system in transgenic plants. Plant J. *11*:605-612.

4. Armaleo, D., G.N. Ye, T.M. Klein, K.B. Shark, J.C. Sanford, and S.A. Johnston. 1990. Biolistic nuclear transformation of *Saccharomyces cerevisiae* and other fungi. Curr. Genet. *17*:97-104.

5. Baker, B., P. Zambryski, B. Staskawicz, and S.P. Dinesh-Kumar. 1997. Signaling in plant-microbe interactions. Science *276*:726-733.

6. Barton, K., A. Binns, A. Matzke, and M.-D. Chilton. 1983. Regeneration of intact tobacco plants containing full length copies of gentically engineered T-DNA, and transmission of T-DNA to R1 progeny. Cell *32*:1033-1043.

7. Bechtold, N., J. Ellis, and G. Pelletier. 1993. In planta *Agrobacterium* mediated gene transfer by infiltration of adult *Arabidopsis thaliana* plants. C. R. Acad. Sci. III *316*:1194-1199.

8. Bent, A.F., B.N. Kunkel, D. Dahlbeck, K.L. Brown, R. Schmidt, J. Giraudat, J. Leung, and B. Staskawicz. 1994. RSP2 of *Arabidopsis thaliana*: a leucine-rich repeat class of plant disease resistance genes. Science *265*:1855-1860.

9. Bevan, M. 1984. Binary *Agrobacterium* vectors for plant transformation. Nucleic Acids Res. *12*:8711-8721.

10. Bevan, M.W., R.B. Flavell, and M.-D. Chilton. 1983. A chimeric antibiotic resistance gene as a selectable marker for plant cell transformation. Nature *304*:184-187.

11. Binns. A.N. 1990. *Agrobacterium*-mediate gene delivery and the biology of host range limitations. Physiol. Plant. *79*:135-139.

12. Binns, A.N. and M.F. Thomashow. 1988. Cell biology of *Agrobacterium* infection and transformation of plants. Annu. Rev. Microbiol. *42*:575-606.

13. Birch, R.G. 1997. Plant transformation: problems and strategies for practical application. Ann. Rev. Plant Physiol. Plant Mol. Biol. *48*:297-326.

14. Boynton, J.E., N.W. Gillham, E.H. Harris, J.P. Hosier, A.M. Johnson, A.R. Jones, B.L. Randolph-Anderson, K. Robertson, T.M. Klein, K.B. Shark, and J.C. Sanford. 1988. Chloroplast transformation in *Chlamydomonas* with high velocity microprojectiles. Science *240*:1534-1536.

15. Brisson, N., J. Faszkowski, J.R. Penswick, B. Gronenborn, I. Potrykus, and T. Hohn. 1984. Expression of a bacterial gene in plants using a viral vector. Nature *310*:511-514.

16. Budar, F., L. Thia-Toong, M. van Montagu, and J.P. Hernalsteens. 1986. *Agrobacterium*-mediated gene transfer results mainly in transgenic plants transmitting T-DNA as a single Mendelian factor. Genetics *114*:303-313.

17. Butow, R.A. and T.D. Fox. 1990. Organelle transformation: shoot first, ask questions later. Trends Biotech. *15*:465-471.

18. Caddick, M.X., A.J. Greenland, I. Jepson, K.-P. Krause, N. Qu, K.V. Riddell, M.G. Salter, W. Schuch, U. Sonnewald, and A.B. Tomsett. 1998. An ethanol inducible gene switch for plants used to manipulate carbon metabolism. Nat. Biotechnol. *16*:177-180.

19. Cangelosi, G.A., R.G. Ankenbauer, and E.W. Nester. 1990. Sugars induce the *Agrobacterium* virulence genes through a periplasmic binding protein and a transmembrane signal protein. Proc. Natl. Acad. Sci. USA *87*:6708-6712.

20. Chalfie, M., Y. Tu, G. Euskirchen, W.W. Ward, and D.C. Prasher. 1994. Green fluorescent protein as a marker for gene expression. Science *263*:802-805.

21. Chang, S.-S., K.-K. Park, B.C. Kim, B.J. Kang, D.U. Kim, and H.-G. Nam. 1995. Stable genetic transformation of *Arabidopsis thaliana* by *Agrobacterium* inoculation in planta. Plant J. *5*:551-558.

22. Charest, P.J., J. Hattori, J. De Moor, V.N. Iyer, and B.L. Miki. 1990. *In vitro* study of transgenic tobacco expressing *Arabidopsis* wild type and mutant acetohydroxyacid synthase genes. Plant Cell Rep. *8*:643-647.

23. Chilton, M.-D. 1993. *Agrobacterium* gene transfer: progress on a "poor man's vector" for maize. Proc. Natl. Acad. Sci. USA *90*:3119-3120.

24. Chilton, M.-D., M.H. Drummond, D.J. Merlo, D. Sciaky, A.L. Montoya, M.P. Gordon, and E.W. Nester. 1977. Stable incorporation of plasmid DNA into higher plant cells: molecular basis of crown gall tumorigenesis. Cell *11*:263-271.

25. Christen, A.A., M.A. Kirkpatrick, and P.F. Lurquin. 1984. Antibiotics and bacterial auxotrophic

mutants in the co-cultivation of plant protoplasts and bacterial cells or spheroplasts. Z. Pflanzen-physiol. Bd. *113*:213-221.

26. Christie, P.J. 1997. *Agrobacterium tumefaciens* T-complex transport apparatus: a paradigm for a new family of multifunctional transporters in eubacteria. J. Bacteriol. *179*:3085-3094.

27. Clough, S.J. and A.F. Bent. 1998. Floral dip: a simplified method for *Agrobacterium*-mediated transformation of *Arabidopsis thaliana*. Plant J. *16*:735-743.

28. Covacci, A., J.L. Telford, G. Del Giudice, J. Parsonnet, and R. Rappuoli. 1999. *Helicobacter pylori* virulence and genetic geography. Science *284*:1328-1333.

29. Crossway, A., J.V. Oakes, J.M. Irvine, B. Ward, V.C. Knauf, and C.K. Shewmaker. 1986. Integration of foreign DNA following microinjection of tobacco mesophyll protoplasts. Mol. Gen. Genet. *202*:179-185.

30. Czakó, M. and L. Márton. 1986. Independent integration and seed-transmission of the T_R-DNA of the octopine Ti plasmid pTi Ach5 in *Nicotiana plumbaginifolia*. Plant Mol. Biol. *6*:101-109.

31. Daniell, H., J. Vivekananda, B.L. Nielsen, G.N. Ye, K.K. Tervari, and J.C. Sanford. 1990. Transient foreign gene expression in chloroplasts in cultured tobacco cells after biolistic delivery of chloroplast vectors. Proc. Natl. Acad. Sci. USA *87*:88-92.

32. Deblaere, R., A. Reynaerts, H. Hofte, J.P. Hernalsteens, J. Leemans, and M. Van Montagu. 1987. Vectors for cloning in plant cells. Methods Enzymol. *153*:277-292.

33. De Block, M., D. De Brouwer, and P. Tenning. 1989. Transformation of *Brassica napus* and *Brassica oleracea* using *Agrobacterium tumefaciens* and the expression of the *bar* and *neo* genes in the transgenic plants. Plant Physiol. *91*:694-703.

34. De Framond, A.J., W.E. Back, W.S. Chilton, L. Kayes, and M.-D. Chilton. 1986. Two unlinked T-DNAs can transform the same tobacco plant cell and segregate in the F1 generation. Mol. Gen. Genet. *202*:125-131.

35. Deroles, S.C. and R.C. Gardner. 1988. Analysis of the T-DNA structure in a large number of transgenic petunias generated by *Agrobacterium*-mediated transformation. Plant Mol. Biol. *11*:365-372.

36. Dessaux, Y., A. Petit, and J. Tempé. 1992. Opines in *Agrobacterium* biology, p. 109-136. *In* D.P. Verma (Ed.), Molecular Signalling in Plant-Microbe Communication. CRC Press, Boca Raton, FL.

37. Faiss, M., J. Zalubilová, M. Strnad, and T. Schmülling. 1997. Conditional transgenic expression of the *ipt* gene indicates a function for cytokinins in paracrine signaling in whole tobacco plants. Plant J. *12*:401-415.

38. Feldmann, K.A. and M.D. Marks. 1987. *Agrobacterium*-mediated transformation of germinating seeds of *Arabidopsis thaliana*: a non-tissue culture approach. Mol. Gen. Genet. *208*:1-9.

39. Feldmann, K.A., M.D. Marks, M.L. Christianson, and R.S. Quatrano. 1989. A dwarf mutant of *Arabidopsis* generated by T-DNA insertional mutagenesis. Science *243*:1351-1354.

40. Feldmann, K.A., T.J. Carson, S.A. Coomber, C.E. Farrance, M.A. Mandel, and A.M. Wierzbicki. 1990. T-DNA insertional mutagenesis in *Arabidopsis thaliana*, p. 109-120. *In* Horticultural Biotechnology. Wiley-Liss, New York.

41. Fillati, J.A., J. Sellmer, B. McCown, B. Hassig, and L. Comai. 1987. *Agrobacterium*-mediated transformation and regeneration of *Populus*. Mol. Gen. Genet. *206*:192-200.

42. Fitch, M.M.M., R.M. Manshardt, D. Gonsalves, J.L. Slighton, and J.C. Sanford. 1990. Stable transformation of papaya via microprojectile bombardment. Plant Cell Rep. *9*:189-195.

43. Fox, T.D., J.C. Sanford, and T.W. McMullin. 1988. Plasmids can stably transform yeast mitochondria lacking endogenous mt DNA. Proc. Natl. Acad. Sci. USA *85*:7288-7292.

44. Fraley, R.T., S.L. Dellaporta, and D. Papahadjopomlos. 1982. Liposome-mediated delivery of tobacco mosaic virus RNA into tobacco protoplasts: a sensitive assay for monitoring liposome-protoplast interactions. Proc. Natl. Acad. Sci. USA *79*:1859-1863.

45. Fraley, R.T., S.G. Rogers, R.B. Horsch, P.R. Sanders, J.S. Flick, S.P. Adams, M.L. Bittner, L.A. Band, C.L. Fink, Y.S. Fry, et al. 1983. Expression of bacterial genes in plant cells. Proc. Natl. Acad. Sci. USA *80*:4803-4807.

46. Fromm, M.E., L.P. Taylor, and V. Walbot. 1986. Stable transformation of maize after gene transfer by electroporation. Nature *319*:791-793.

47. Fromm, M.E., F. Morrish, C. Armstrong, R. Williams, J. Thomas, and T.M. Klein. 1990. Inheritance and expression of chimeric genes in the progeny of transgenic maize plants. BioTechnology *8*:833-838.

48. Fry, J., A. Barnason, and R.B. Horsch. 1987. Transformation of *Brassica napus* with *Agrobacterium*

tumefaciens based vectors. Plant Cell Rep. *6*:321-330.

49. Gamborg, O.L., R.A. Miller, and K. Ojima. 1968. Nutrient requirements of suspension cultures of soybean root cells. Exp. Cell Res. *50*:151-158.

50. Gasser, C.S. and R.T. Fralcy. 1989. Genetically engineering plants for crop improvement. Science *244*:1293-1299.

51. Gatz, C., C. Frohberg, and R. Wendenburg. 1992. Stringent repression and homogeneous de-repression by tetracycline of a modified CaMV 35S promoter in intact transgenic plants. Plant J. *2*:397-404.

52. Goff, S.A., T.M. Klein, B.A. Roth, M.E. Fromm, K.C. Cone, J.P. Radicella, and V.L. Chandler. 1990. Transactivation of anthocyanin biosynthetic genes following transfer of B regulatory genes into maize tissue. EMBO J. *9*:2517-2522.

53. Goodman, R.M., H. Hauptli, A. Crossway, and V.C. Knauf. 1987. Gene transfer in crop improvement. Science *236*:48-54.

54. Grant, S.R. 1999. Dissecting the mechanisms of posttranscriptional gene silencing: divide and conquer. Cell *96*:303-306.

55. Gruber, M.Y and W.L. Crosby. 1993. Vectors for plant transformation. p. 89-119. *In* B.R. Glick and J.E. Thompson (Eds.), Methods in Plant Molecular Biology and Biotechnology. CRC Press, Boca Raton, FL.

56. Guerincau, F., L. Brooks, J. Meadows, A. Lucy, C. Robinson, and P. Mullineaux. 1990. Sulfonamide resistance gene for plant transformation. Plant Mol. Biol. *15*:127-134.

57. Hamill, J.D., A.J. Parr, M.J.C. Rhodes, R.J. Robins, and N.J. Walton. 1987. New routes to plant secondary products. BioTechnology *5*:800-804.

58. Hanley-Bowdoin, L., J.S. Elmer, and S.G. Rogers. 1988. Transient expression of heterologous RNAs using tomato golden mosaic virus. Nucleic Acids Res. *16*:10511-10528.

59. Haseloff, J. and B. Amos. 1995. GFP in plants. Trends Genet. *11*:328-329.

60. Hansen, G. and M.-D. Chilton. 1996. "Agrolistic" transformation of plant cells: integration of T-stands generated in planta. Proc. Natl. Acad. Sci. USA *93*:14978-14983.

61. Hansen, G., R.D. Shillito, and M.-D. Chilton. 1997. T-strand integration in maize protoplasts after codelivery of a T-DNA substrate and virulence genes. Proc. Natl. Acad. Sci. USA *94*:11726-11730.

62. Hayes, R.J., I.T.D. Petty, R.H.A. Coutts, and K.W. Buck. 1988. Gene amphilication and expression in plants by a replicating geminivirus vector. Nature *334*:179-182.

63. Herreta-Estrella, L. and J. Simpson. 1988. Foreign gene expression in plants. p. 131-160. *In* C.H. Shaw (Ed.), Plant Molecular Biology: A Practical Approach. IRL Press, Washington, D.C.

64. Hoekema, A., M. van Haaren, A. Fellinger, P. Hooykaas, and R. Schilperoort. 1985. Non-oncogenic plant vectors for use in the *Agrobacterium* binary system. Plant Mol. Biol. *5*:85-89.

65. Hoekema, A., P.R. Hirsch, P.J.J. Hooykaas, and R.A. Schilperoort. 1983. A binary plant vector strategy based on separation of *vir-* and T-region of the *Agrobacterium tumefaciens* Ti-plasmid. Nature *303*:179-180.

66. Hood, E.E., G.L. Helmer, R.T. Fraley, and M.-D. Chilton. 1986. The hypervirulence of *Agrobacterium* A281 is encoded in a region of pTiBo542 outside the T-DNA. J. Bacteriol. *178*:1291-1298.

67. Hood, E.E., S.B. Gelvin, L.S. Melchers, and A. Hoekema. 1993. New *Agrobacterium* helper plasmids for gene transfer to plants. Transgenic Res. *2*:208-218.

68. Hooykaas, P.J.J. and R.A. Schilperoort. 1992. *Agrobacterium* and plant genetic engineering. Plant Mol. Biol. *19*:15-38.

69. Horsch, A., J.E. Fry, N.L. Hoffman, D. Eicholtz, S.D. Rogers, and R.T. Fraley. 1985. A simple and general method for tranferring genes into plants. Science *227*:1229-1231.

70. Jefferson, R.A. 1987. Assaying chimeric genes in plants: the GUS gene fusion system. Plant Mol. Biol. Rep. *5*:387-398.

71. Jefferson, R.A., T.A. Kavanagh, and M.W. Bevan. 1987. GUS fusions: β-glucuronidase as a sensitive and versatile gene fusion marker in higher plants. EMBO J. *6*:3901-3911.

72. Johnston, S.A. 1990. Biolistic transformation: microbes to mice. Nature *346*:776-777.

73. Johnston, S.A., P.Q. Anzsiano, K. Shark, J.C. Sanford, and R.A. Butow. 1988. Mitochondria transformation in yeast by bombardment with microprojectiles. Science *240*:1538-1542.

74. Jones, A.M., K.-H. Im, M.A. Savka, M.-J. Wu, N.G. DeWitt, R. Shillito, and A.N. Binns. 1998. Auxin-dependent cell expansion mediated by overexpressed auxin-binding protein 1. Science

282:1114-1117.

75. Jorgensen, R., C. Snyder, and J.D.G. Jones. 1987. T-DNA is organized predominantly in inverted repeat structures in plants transformed with *Agrobacterium tumefaciens* C58 derivatives. Mol. Gen. Genet. *207*:471-479.

76. Klee, H., M. Yanofsky, and E. Nester. 1985. Vectors for the transformation of higher plants. BioTechnology *3*:637-642.

77. Klein, T.M., E.C. Harper, Z. Svab, J.C. Sanford, M.E. Fromm, and P. Maliga. 1988. Stable genetic transformation of intact *Nicotiana* cells by the particle bombardment process. Proc. Natl. Acad. Sci. USA *85*:8502-8505.

78. Klein, T.M., E.D. Wolf, R. Wu, and J.C. Sanford. 1987. High-velocity microprojectiles for delivering nucleic acids into living cells. Nature *327*:70-73.

79. Koncz, C. and J. Schell. 1986. The promoter of TL-DNA gene 5 controls the tissue-specific expression of chimearic genes carried by a novel type of *Agrobacterium* binary vector. Mol. Gen. Genet. *204*:383-396.

80. Koncz, C., O. Olsson, W.H.R. Langridge, J. Schell, and A.A. Szalay. 1987. Expression and assembly of functional bacterial luciferase in plants. Proc. Natl. Acad. Sci. USA *84*:131-135.

81. Larkin, P.J. and W.R. Scowcroft. 1982. Somaclonal variation—a novel source of variability from cell cultures for plant improvement. Theor. Appl. Genet. *60*:197-214.

82. Li, Y., G. Hagen, and T.J. Guilfoyle. 1992. Altered morphology in transgenic tobacco plants that overproduce cytokinins in specific tissues and organs. Dev. Biol. *153*:386-395.

83. Lloyd, A.M., A.R. Barnason, S.G. Rogers, M. Byrne, F.T. Fraley, and R. Horsch. 1986. Transformation of *Arabidopsis thaliana* with *Agrobacterium tumefaciens*. Science *234*:464-466.

84. Ma, H., M.F. Yanofsky, H.J. Klee, J.L. Bowman, and E.M. Meyerowitz. 1992. Vectors for plant transformation and cosmid libraries. Gene *117*:161-167.

85. McCown, B.H., D.E. McCabe, D.R. Russell, D.J. Robinson, K.A. Barton, and K.F. Raffa. 1991. Stable transformation of *Populus* and incorporation of pest resistance by electric discharge particle acceleration. Plant Cell Rep. *9*:590-596.

86. McKenzie, M.J., V. Mett, P.H.S. Reynolds, and P.E. Jameson. 1998. Controlled cytokinin production in transgenic tobacco using a copper-inducible promoter. Plant Physiol. *116*:969-977.

87. Mett, V.L., L.P. Lochhead, and P.H.S. Reynolds. 1993. Copper-controllable gene expression system for whole plants. Proc. Natl. Acad. Sci. USA 90:4567-4571.

88. Mett, V.L., E. Podivinsky, A.M. Tennant, L.P. Lochhead, W.T. Jones, and P.H.S. Reynolds. 1996. A system for tissue-specific copper controllable gene expression in transgenic plants: nodule-specific antisense of aspartate aminotransferase-P2. Transgenic Res. 5:105-113.

89. Miranda, A., G. Jansenn, L. Hodges, E. G. Peralta, and W. Ream. 1992. *Agrobacterium tumefaciens* transfers extremely long T-DNAs by a unidirectional mechanism. J. Bacteriol. *174*:2288-2297.

90. Moffat, A.S. 1998. Toting up the early harvest of transgenic plants. Science *282*:2176-2178.

91. Murashige, T. and F. Skoog. 1962. A revised medium for rapid growth and bioassays with tobacco tissue cultures. Physiol. Plant *15*:473-497.

92. Oger, P., A. Petit, and Y. Dessaux. 1997. Genetically engineered plants producing opines alter their biological environment. Nat. Biotechnol. *15*:369-372.

93. Ow, D.W., K.V. Wood, M. De Luca, J.R. De Wet, D.R. Helinski, and S.H. Howell. 1986. Transient and stable expression of the firefly luciferase gene in plant cells and transgenic plants. Science *234*:856-858.

94. Pang, S.-Z., D.L. DeBoer, Y. Wan, G. Ye, J.G. Layton, M.K. Neher, C.L. Armstrong, J.E. Fry, M.A.W. Hinchee, and M.E. Fromm. 1996. An improved green fluorescent protein gene as a vital marker in plants. Plant Physiol. *112*:893-900.

95. Paszkowski, J. and M.W. Saul. 1986. Direct gene transfer to plants. Methods Enzymol. 118:668-684.

96. Phelep, M., A. Petit, L. Martin, E. Duhoux, and J. Tempé. 1991. Transformation and regeneration of a nitrogen-fixing tree, *Allocasuarina verticillata* Lam. BioTechnology 9:461466.

97. Pythoud, F., V.P. Sinkar, E.W. Nester, and M.P. Gordon. 1987. Increased virulence of *Agrobacterium rhizogenes* conferred by the *vir* region of pTiBo542: application to genetic engineering of poplar. BioTechnology 5:1323-1327.

98. Redig, P., T. Schmülling, and H. van Onckelen. 1996. Analysis of cytokinin metabolism in *ipt* transgenic tobacco by liquid chromatography-tandem mass spectrometry. Plant Physiol. *122*:141-

148.

99. Robertson, D., A.K. Weissinger, R. Ackley, S. Glover, and R.R. Sederoff. 1992. Transient expression and stable transformation of Norway spruce (*Picea abies* L. Karst) somatic embryos using microprojectile bombardment. Plant Mol. Biol. *19*:925-934.

100. Röder, F.T., T. Schmülling, and C. Gatz. 1994. Efficiency of the tetracycline-dependent gene expression system: complete suppression and efficient induction of the *rolB* phenotype in transgenic plants. Mol. Gen. Genet. *243*:32-38.

101. Rogers, S.G., L.A. Brand, S.B. Holder, E.S. Sharp, and M.J. Brackin. 1983. Amplification of the *aroA* gene from *E. coli* results in tolerance to the herbicide glyphosate. Appl. Environ. Microbiol. *46*:37-43.

102. Sautter, C., H. Waldner, C. Neuhaus-Url, A. Galli, G. Neuhaus, and I. Portykus. 1991. Microtargeting high efficiency gene transfer using a novel approach for the acceleration of micro-projectiles. BioTechnology *9*:1080-1084.

103. Savka, M.A. 1993. Validity of the opine concept in plant-bacterial interactions [Ph.D. Dissertation]. University of Illinois at Urbana-Champaign, Urbana, IL.

104. Savka, M.A. and S.K. Farrand. 1997. Modification of rhizobacterial populations by engineering bacterial utilization of a novel plant-produced resource. Nat. Biotechnol. *15*:363-368.

105. Scott, A., S. Wyatt, P.-L. Tsou, D. Robertson, and N. Strömgren Allen. 1999. Model system for plant cell biology: GFP imaging in living onion epidermal cells. BioTechniques *26*:1125-1132.

106. Shah, D.M., R.B. Horsch, H.J. Klee, G.M. Kishmore, J.A. Winter, N.E. Tumer, C.M. Hironaka, P.R. Sanders, C.S. Gasser, S. Aykent, et al. 1986. Engineering herbicide tolerance in transgenic plants. Science *233*:478-480.

107. Sheerman, S. and M.W. Bevan. 1988. A rapid transformation method for *Solanum tuberosum* using binary *Agrobacterium tumefaciens* vectors. Plant Cell Rep. *7*:13-21.

108. Sheikholeslam, S.N. and D.P. Weeks. 1987. Acetosyringone promotes high efficiency transformation of *Arabidopsis thaliana* explants by *Agrobacterium tumefaciens*. Plant Mol. Biol. *8*:291-298.

109. Shillito, R.D., M.W. Saul, J. Paszkowski, M. Muller, and I. Potrykus. 1985. High efficiency direct gene transfer to plants. BioTechnology *3*:1099-1104.

110. Southern, E.M. 1975. Detection of specific sequences among DNA fragments separated by gel electrophoresis. J. Mol. Biol. *98*:503-517.

111. Stachel, S.E., E. Messens, M. van Montagu, and P. Zambryski. 1985. Identification of the signal molecules produced by wounded plant cells that activate T-DNA transfer in *Agrobacterium tumefaciens*. Nature *318*:624-629.

112. Stachel, S.E., G. An, C. Flores, and E.W. Nester. 1985. A Tn3 *lacZ* transposon for the random generation of β-galactosidase gene fusions: application to the analysis of gene expression in *Agrobacterium*. EMBO J. *4*:891-898.

113. Taylor, B.H., J.R. Manhart, and R.M. Amasino. 1993. Isolation and characterization of plant DNAs, p. 37-47. *In* B.R. Glick and J.E. Thompson (Eds.), Methods in Plant Molecular Biology and Biotechnology. CRC Press, Boca Raton, FL.

114. Tepfer, D. 1984. Transformation of several species of higher plants by *Agrobacterium rhizogenes*: sexual transmission of the transformed genotype and phenotype. Cell *37*:959-967.

115. Tepfer, D., L. Metzger, and R. Prost. 1989. Use of roots transformed by *Agrobacterium rhizogenes* in rhizosphere research: applications in studies of cadmium assimilation from sewage sludges. Plant Mol. Biol. *13*:295-302.

116. Titomirov, A.V. and A.V. Zelenin. 1989. New methods in transinfection of mammalian cells: a minireview. Mol. Biol. *359*:1155-1167.

117. Valvekens, D., M. Van Montagu, and M. Van Lusebettens. 1988. *Agrobacterium tumefaciens*-mediated transformation of *Arabidopsis thaliana* root explants by using kanamycin selection. Proc. Natl. Acad. Sci. USA *85*:5536-5540.

118. Vanden Elzen, P.J.M., J. Townsend, K.Y. Lee, and J.R. Bedbrook. 1985. A chimeric hygromycin resistance gene as a selectable marker in plant cells. Plant Mol. Biol. *5*:299-306.

119. Vaucheret, H., C. Beclin, T. Elmayan, F. Feuerbach, C. Godon, J.B. Morel, P. Mourrain, J.C. Palauqui, and S. Vernhettes. 1998. Transgene-induced gene silencing in plants. Plant J. *16*:651-659.

120. Wang, E.R., L. Herrera-Estrella, M. Van Montagu, and P. Zambryski. 1984. Right 25 bp terminus sequences of the nopaline T-DNA is essential for and determines direction of DNA transfer from *Agrobacterium* to the plant genome. Cell *38*:35-41.

121.Winans, S.C., D.L. Burns, and P.J. Christie. 1996. Adaptation of a conjugal transfer system for the export of pathogenic macromolecules. Trends Microbiol. *4*:64-68.

122.Yadav, N.S., J. Vanderlayden, D.R. Bennett, W.M. Barnes, and M.-D. Chilton. 1982. Short direct repeats flank the T-DNA on a nopaline Ti plasmid. Proc. Natl. Acad. Sci. USA *79*:6322-6326.

123.Yang, N.S., J. Burkholder, B. Roberts, B. Martinell, and D. McCabe. 1990. In vitro and in vivo gene transfer to mammalian somatic cells by particle bombardment. Proc. Natl. Acad. Sci. USA *87*:9568-9572.

124.Zambryski, P.C. 1988. Basic processes underlying *Agrobacterium*-mediated DNA transfer to plant cells. Annu. Rev. Genet. *22*:1-30.

125.Zambryski, P.C. 1992. Chronicles from the *Agrobacterium*-plant cell DNA transfer story. Annu. Rev. Plant Physiol. Plant Mol. Biol. *43*:465-490.

126.Zambryski, P., H. Joos, C. Genetello, J. Leemans, M. Van Montagu, and J. Schell. 1983. Ti plasmid vector for the introduction of DNA into plant cells without alteration of their normal regeneration capacity. EMBO J. *2*:2143-2150.

127.Zupan, J.R. and P. Zambryski. 1995. Tranfer of T-DNA from *Agrobacterium* to the plant cell. Plant Physiol. *107*:1041-1047.

Introduction of Therapeutic cDNAs into the Lungs of Humans and Animals

Larry G. Johnson
*Division of Pulmonary and Critical Care Medicine,
Department of Medicine, The Cystic Fibrosis/Pulmonary
Research and Treatment Center, The University of North
Carolina at Chapel Hill, Chapel Hill, NC, USA*

INTRODUCTION

The identification of human genes linked to clinical disease from human genome research has raised hopes for the development of specific genetic therapies for many inherited and acquired diseases. Cystic fibrosis (CF) and alpha-1 antitrypsin (α_1-AT) deficiency are the most common inherited disorders affecting the lung that are potential targets for gene transfer, whereas lung cancer is the most common acquired disorder that serves as a potential target. Primary pulmonary hypertension has also recently become a disease target for gene transfer.

In cystic fibrosis, the site where CF lung disease begins is still debated, with both the superficial columnar epithelial cells lining the lumen of the small airways and the serous cells of submucosal glands having been identified as potential sites. Clinical data tends to support the theory that the disease begins in the small airways (13,156). Where the disease begins is relevant since luminal (airway) delivery of gene transfer vectors targets the superficial columnar airway epithelium, while intravenous (blood) delivery may be required to target the submucosal glands and basal cells in the airway. Basal cells are potential targets because of their ability to differentiate into columnar epithelial cells (64).

Gene Transfer Methods
Edited by Pamela A. Norton and Laura F. Steel
©2000 Eaton Publishing, Natick, MA

Following initial in vitro complementation studies (30,34,79,118,135, 138,188,192), investigators rapidly moved to clinical safety and efficacy trials of gene transfer vectors delivered by luminal application to the airways of CF patients. While some evidence for gene transfer was detected, the efficiency and efficacy of gene transfer failed to meet expectations and did not fully correct the known biochemical defects ascribed to this disorder (2,9,18,22, 57,70,91,116,132,162,185,187,191). This failure forced exploration of potential barriers to gene transfer in airways and stimulated efforts to develop strategies to overcome these barriers.

Although an ongoing clinical gene transfer safety and efficacy study has been initiated in respiratory epithelia of α_1-AT deficient patients, the hepatocyte may be the best cellular target for gene therapy of this disorder. α_1-AT deficiency is the most common genetic cause of emphysema and a common cause of cirrhosis in children (115,125,175). α_1-AT is a serine protease inhibitor produced in the liver and a member of the serine protease inhibitor family known as the serpins. This secreted protein enters the lung via the systemic circulation and is the major inhibitor of neutrophil elastase in the lung. The most common mutant form of this disease (the protease inhibitor Z or PiZ genotype) is associated with inadequate levels of circulating protein and accumulation of mutant protein in the liver that can lead to cirrhosis. Because α_1-AT is a secretory protein, gene therapy approaches could target a variety of organs.

Lung cancer is the most common acquired disease target and the most complex target for gene therapy. Unlike monogenic disorders, mutations in multiple somatic genes may be the cause of the disorder. Furthermore, strategies may need to be developed that target both local disease and systemic metastasis. Although the molecular pathogenesis has not been entirely delineated, these somatic mutations mediate active immunosuppression and improperly regulate cellular growth. Strategies for gene therapy of lung cancer must therefore address multiple potential mechanisms of tumorigenesis (32,53). These include tumor suppressor gene replacement therapy for mutations in genes such as p53 that are known to occur in approximately 60% of non-small cell lung cancers (NSCLC), antisense therapies for dominant oncogenes, e.g., k-ras, associated with lung cancer, toxin gene/prodrug (HSVtk/gancyclovir or cytosine deaminase/5-FC) or suicide gene therapy approaches, and immuno-gene therapy with production of cytokines (e.g., IL-7, IL-12, and GM-CSF) (32,53).

Pulmonary hypertension may be a potential disease target for gene therapy (16,56,74). Pulmonary hypertension is a disease of the pulmonary vasculature that occurs when damage to the endothelium leads to uncontrolled vascular remodeling, increased vascular pressures and resistance, right-sided heart failure, and death. Primary pulmonary hypertension (PPH) is the prototypical disease, although a variety of secondary causes are known (137). PPH has a female-to-male preponderance, with patients most commonly presenting in the third and fourth decades. The predominant presentation is dyspnea of insidious onset in an otherwise healthy person. Mean survival is only 2 to 3 years in patients presenting with significant functional impairment (55,137).

PPH is characterized by uncontrolled pulmonary vascular remodeling due to endothelial proliferation and has been associated with increased production of endothelin and other vasoconstrictive mediators and decreased production of endothelial nitric oxide synthase (eNOS) (39,90,161). A recent human study suggests that endothelial cell prostacyclin production is also reduced in individuals with severe pulmonary hypertension (158). Intravenous prostacyclin has been approved as a treatment of primary pulmonary hypertension for patients who have moderate to severe functional impairment that is unresponsive to conventional therapy. Clinical trials of prostacyclin have demonstrated objective improvements in dyspnea and exercise tolerance and a reduction in mortality (7). The side effects of prostacyclin, including flushing, jaw pain, and diarrhea, are generally well-tolerated. However, placement of a permanent central venous catheter for intravenous delivery is associated with catheter-related complications.

Overexpression in transgenic mice of the rat prostacyclin synthase (PGIS) cDNA under the control of an epithelial specific surfactant protein C (SP-C) promoter has been shown to prevent chronic hypoxic pulmonary hypertension (56). Adenoviral (Ad)-mediated overexpression of constitutive endothelial nitric oxide synthase (ceNOS) in alveolar epithelia and in medium and small pulmonary vessels of rat lung following aerosol administration has been shown to attenuate acute hypoxic pulmonary vasoconstriction, suggesting that luminal gene transfer approaches are feasible (74). Recently, genetically modified smooth muscle cells expressing human eNOS injected into the superior vena cava have also been shown to attenuate monocrotaline-induced pulmonary hypertension (16). Thus, it may also be feasible to deliver prostacyclin synthase and eNOS to lung epithelia or to deliver eNOS ex vivo as potential approaches to gene therapy of PPH.

In this chapter, we review the basic principles of gene transfer vectors used for lung gene transfer, review methods for in vivo delivery, discuss the barriers that have become apparent as a result of in vivo animal and human studies, and explore new strategies to overcome the barriers to gene transfer.

VECTORS FOR LUNG GENE TRANSFER

Naturally occurring viruses known to cause human disease introduce their own viral nucleic acid (DNA or RNA) into the host cell nucleus leading to expression of viral genes that promote viral replication. This unique ability of naturally occurring viruses has been adapted to develop viral vectors that introduce and express therapeutic genes (cDNAs) in lieu of their viral structural genes, which have been deleted. Adenoviruses (Ad), adeno-associated viruses (AAV), and retroviruses each have been altered to introduce therapeutic cDNAs. Ad, AAV, and retroviruses have already been used in clinical trials in CF and lung cancer patients (9,22,70,91,139,140,155,187,191). However, viral vectors can potentially induce a variety of immune and other adverse reactions that have made nonviral vectors an attractive alternative approach. Of the nonviral methods, only cationic liposomes have undergone sufficient investigation to enter initial clinical gene transfer safety and efficacy trials in the lung (2,18,57,116, 132,185). The characteristics of each these vectors are summarized in Table 1.

Viral Vectors

Adenoviral Vectors

Adenoviruses are double-stranded DNA viruses, of which human serotypes 2 and 5 (90% homology) provide the backbone for current Ad vectors (11,58,94,136). Wild-type adenoviruses have a 36-kbp genome consisting of a series of early genes that are responsible for virus replication, antigen presentation and surveillance, and a series of late genes that encode viral structural proteins (11,58,136). Several generations of Ad vectors have now been developed (Figure 1). First generation vectors based on adenovirus serotype 5 (Ad5) have had the early region one (E1) genes deleted (11,37) to make the vector replication-defective, while deletion of the E3 region creates sufficient room to insert therapeutic cDNAs with a suitable promoter (Figure 1A). First generation Ad vectors based on serotype 2 (Ad2) have only had the E1 gene deleted such that the vector genome may actually exceed the size of the wild-type virus genome (136), leading to less efficient packaging of Ad-DNA into viral particles and lower infectious titers (Figure 1A).

E1-deleted second generation vectors based on Ad5 have been developed in which the E2a region has also been mutated to form a temperature-sensitive mutant virus (Figure 1B). This mutant replicates in 293 cells at 32°C, but not at 39°C, potentially bringing an additional safety feature to this vector (35,38, 59,181). E1-deleted second generation vectors have also been developed with deletions in most (except for open reading frame 6, ORF6) or all of the E4 region (Figure 1B) in an attempt to limit late viral gene expression (6,82,168).

Third generation vectors have also been developed with deletions of all viral genes, retaining only a small packaging signal and the inverted terminal repeats (Figure 1C) (46,68,69,92,100,112,122). These vectors are attractive because they can accept insert cDNAs or even genomic DNA in excess of 30 kb (94). For adequate packaging of vector constructs containing smaller cDNAs,

Table 1. Characteristics of Gene Transfer Vectors

Gene Transfer Vectors	cDNA Insert Size (kbp)	Duration of Expression	Transduction of Nondividing Cells	Risk of Insertional Mutagenesis	Induction of Cell-Mediated Immune Response	Induction of Humoral Immune Response
Adenovirus						
1st/2nd generation	7–8	Transient	yes	minimal	yes	yes
High-capacity	>30	long-term?	yes	minimal	unknown[a]	yes
Adeno-associated virus	4.5	long-term	yes	yes	no	yes
Retrovirus						
MuLV[b]	7	long-term	no	yes	no	no
Lentivirus[c]	>7	long-term	yes	yes	no	no
Cationic liposomes	>10	transient	yes	no	rarely	no
Molecular conjugates	>10	transient	yes	no	no	yes

[a]High-capacity adenoviral vectors free of contaminating first-generation he per virus have not been developed to date.

[b]Derived from Moloney MuLV.

[c]Derived from HIV, EIAV, or FIV. Experience with high-capacity adenoviral and lentiviral vectors is limited.

these vectors often require a stuffer sequence (92,112). The absence of viral proteins also decreases the immune response to these vectors. However, these completely deleted vectors require co-infection of producer cells with an Ad helper virus for production of viral structural proteins. Thus, high-capacity Ad vector preparations may be contaminated with first generation Ad helper virus. The advent of Cre-lox recombination techniques has led to a reduction in the amount of Ad helper contamination to as low as 0.5% to 1.0% in most cases. Thus, Ad vectors that have complete deletions of the Ad genome are often referred to as gutless, high-capacity, or helper-dependent Ad vectors.

Ad enters cells by receptor-mediated endocytosis (47). It binds to a high affinity coxsackie and adenoviral receptor (CAR) (10) and is subsequently internalized through an $\alpha_v\beta_{3/5}$ integrin-mediated vesicular (endocytic) process (173). As a result of endosomolytic properties mediated by the Ad penton base, Ad avoids lysosomal degradation such that efficient translocation of the

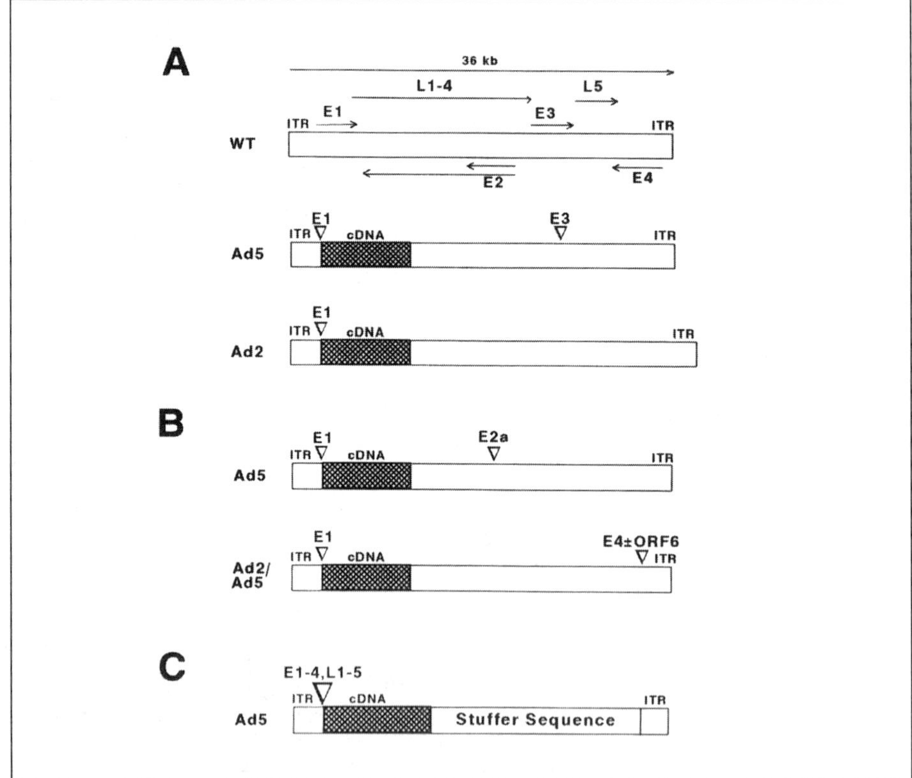

Figure 1. Generations of adenoviral vectors from wild-type adenovirus. (A) First generation vectors. (B) Second generation vectors. (C) Third generation vectors. Ad5, Adenovirus serotype 5 vector; Ad2, adenovirus serotype 2 vector; E, early region viral genes; ITR, inverted terminal repeat; L, late region viral genes; WT, wild-type; ∇, deletion of an early or late region gene; ±, with or without.

DNA to the nucleus occurs where it exists as an episome (extrachromosomal DNA) mediating expression of therapeutic genes (cDNAs). Because Ad vectors do not integrate at high frequencies, transient expression occurs such that repetitive administration is required for genetic disorders, but may not be necessary for other disorders such as lung cancer.

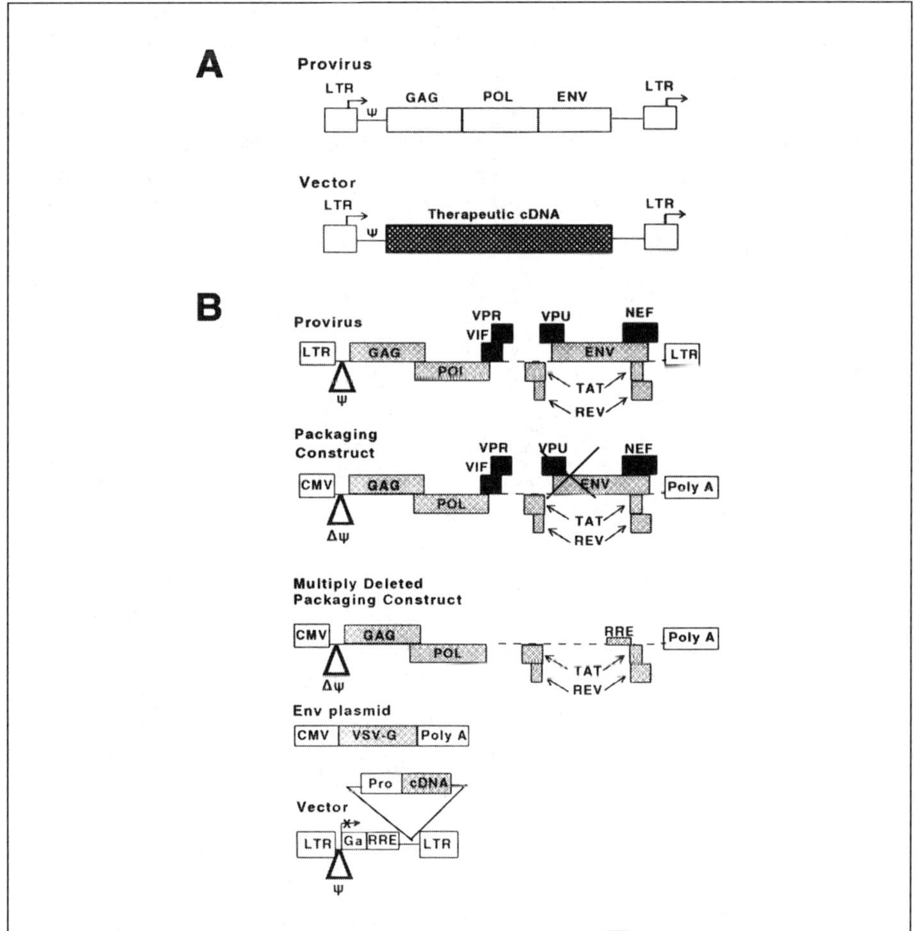

Figure 2. Maps of murine leukemia virus (MuLV)-derived retroviral and human lentiviral (HIV-derived) vectors. (A) Provirus of wild-type murine leukemia retrovirus and a replication-defective retroviral vector. (B) HIV provirus and HIV-based lentiviral packaging plasmids and vectors for production of replication defective vector. HIV vectors are produced by cotransfection of a packaging plasmid, an envelope (Env) plasmid, and a vector into 293 cells, followed by concentration of vector-containing supernatant by ultracentrifugation. Note that the HIV genome contains a number of accessory genes not present in MuLV including *vif, vpr, vpu, nef, rev,* and *tat*. In the multiply deleted packaging construct, several of the accessory genes have been deleted. LTR, long terminal repeat; Ψ, viral packaging signal; ΔΨ, deleted viral packaging signal; VSV-G, vesicular stomatitis virus glycoprotein cDNA; pro, promoter; CMV, cytomegalovirus immediate early promoter.

Adeno-Associated Virus Vectors

AAV vectors are attractive for gene therapy because they offer the possibility of long-term expression with a high degree of safety. They are derived from the naturally defective and nonpathogenic wild-type human parvoviruses, AAV2 and AAV3 (19,48-52,194). AAV requires the presence of a helper virus, e.g., adenovirus or herpes simplex virus, to replicate or to cause a lytic infection. In the absence of a helper virus, AAV integrates into the host cell genome and becomes latent. Upon a subsequent wild-type adenovirus or wild-type herpesvirus infection, the AAV genome can be rescued (excised) from the chromosome to generate a lytic infection (19,48-52,194).

The small AAV genome (approximately 4.7 kbp) consists of the following: *(i)* inverted terminal repeats at the 5′ and 3′ ends of the molecule (19,51), which play a role in replication and are important for integration into the host cell genome; and *(ii)* the viral genes *rep* and *cap*, which mediate viral replication and nucleocapsid formation. Deletion of *rep* and *cap* creates an AAV vector with an insert size of approximately 4.5 kb, which is at the upper size limit for insertion of full-length wild-type CFTR (coding region of approximately 4.5 kb) driven by an exogenous promoter. AAV vectors have generally been produced by transient transfection of AAV vector and AAV helper plasmids into 293 cells in the presence of Ad helper (E1-deleted) virus. Despite cesium chloride banding, this method invariably resulted in Ad contamination of recombinant AAV (rAAV) preparations requiring heat inactivation. Recently, Ad-free helper systems have been designed with transient transfection of AAV vector, Ad helper functions, and AAV *rep* and *cap* functions on separate plasmids to eliminate Ad contamination (178). Another advance has been the use of heparan sulfate affinity columns that improve the titer and yield of rAAV production during purification (154,197).

Retroviral Vectors

Retroviruses have been extensively used in the laboratory for stable expression of foreign cDNAs. Retroviruses derived from murine leukemia virus (MuLV) are RNA viruses whose genomes consist of two viral long terminal repeats (LTRs) that are important for cellular integration, but also contain promoter elements, a packaging signal, and a series of structural genes, *gag*, *pol*, and *env* (Figure 2A). These structural genes encode the capsid protein, reverse transcriptase and an integrase, and the envelope glycoprotein, respectively. Deletion of *gag*, *pol*, and *env* creates space for insertion of therapeutic cDNAs (genes) into the retroviral genome forming a replication defective retroviral vector (110). The genomes of lentiviruses are more complex and encode a variety of accessory proteins and pathogenesis factors that are not present in the simpler genomes of MuLV (113,114,200). Major deletions in these genes have enabled the development of lentiviral vectors for gene transfer (Figure 2B) (113,114,200). Exogenous (internal) promoters may also be included in the sequences of the inserted gene or alternatively, the viral 5′ LTR may be used to drive transcription of the therapeutic gene. Deletions in the LTR to prevent transcription (self-inactivating or SIN vectors) have been recently introduced as a safety feature of MuLV and human lentiviral (such as derived from human immunodeficiency virus [HIV]) vectors (25,199). Replication-defective vectors have traditionally been produced from packaging cell lines that provide the retroviral structural and replication proteins *in trans*. Transient transfection techniques that place vector, helper, and envelope functions on separate plasmids have also been developed to permit rapid production of retroviral vectors for testing in vitro and in vivo (124,149).

The envelope glycoproteins of a wild-type retrovirus bind to cell surface receptors to facilitate entry of the retrovirus into the cytoplasm where the retroviral RNA is reverse transcribed to form a cDNA, the provirus. This provirus is translocated to the nucleus where it integrates into the host cell chromosomes, and through the normal process of DNA transcription, encodes new viral proteins and new viral RNA, which are assembled at the cell surface into new viral particles. Replication-defective retroviral vectors infect cells by a similar mechanism, but unlike wild-type retroviruses, the integrated provirus from a retroviral vector encodes the therapeutic gene, and viral particles are not produced.

Retroviral vectors are attractive as gene transfer vectors since integration of desired cDNAs into the host cell genome offers the possibility of long-term expression. Previously, the low rates of epithelial cell proliferation in normal human airways (99) combined with the relatively low titers of murine leukemia virus-derived vectors have limited the utility of retroviral vectors in lung epithelia in vivo because of their requirement for proliferating cells (111). However, advances in the pseudotyping of retroviral vectors (15) that permit concentration of titers to greater than 10^9 infectious units (IU)/mL combined with the development of lentiviral vectors derived from human immunodeficiency virus (HIV), equine infectious anemia virus (EIAV), and feline immunodeficiency virus (FIV), may soon overcome the limitations in some cell types (113,114,117,167,200).

Nonviral Vectors

Cationic Liposomes

Cationic liposomes and naked plasmid DNA have been the principal nonviral vectors used in clinical trials to date. Cationic liposomes are composed of cationic lipids mixed in varying molar ratios with cholesterol and dio-leylphoshphatidylethanolamine (DOPE), a neutral phospholipid (96,143). Commonly used cationic lipids for gene transfer include N(1-(2,3-di-oleoxy)propyl) N,N,N trimethylammonium (DOTMA), 1,2-dimyristy-loxypropyl-3-dimethylhydroxyethylammonium bromide (DMRIE), or 3β(N-N′,N′-dimethylamino ethane-carbamoyl) cholesterol (DC-Chol), N(1-(2,3-dioleoxy)propyl) N,N,N trimethylammonium methyl sulfate (DOTAP), p-ethyl dimyristoyl phosphatidyl choline (EDMPC) cholesterol, and N^4-sperminine cholesteryl carbamate (GL-67). Cationic liposomes bind to negatively charged plasmid DNA to form DNA–liposome complexes which may, under conditions of excess molar DNA, have a net negative charge. DNA–liposome complexes (also known as lipoplexes) enter cells primarily by endocytosis (98,195), although the mechanism and specificity of binding to the cell surface, given the potential for negatively charged complexes, have not been clearly delineated. Cationic liposomes also promote the release of plasmid DNA from the endosome into the cytoplasm (195). Like adenoviruses, cationic liposomes do not integrate into the host cell genome, so that expression may be lost with cell division. Thus, repetitive administration of lipoplexes may be required for some applications of gene therapy, particularly genetic disorders.

Naked DNA

Naked or plasmid DNA has also been shown to mediate gene transfer to lung epithelia in vitro and in vivo in the absence of liposomes. In a clinical trial of lipid GL-67 in the nasal epithelium of cystic fibrosis patients, plasmid DNA alone was as effective as GL-67/CFTR plasmid DNA lipoplexes, although the restoration of chloride secretion was small in both cases (185).

DNA–Ligand Complexes

When plasmid DNA is linked to polylysine and a receptor ligand, this forms a DNA–ligand complex or molecular conjugate. This complex may then bind to a specific cell surface receptor and enter the cell by receptor-mediated endocytosis. Addition of an inactivated adenovirus to this complex using an antibody bridge results in more efficient gene transfer in vitro by virtue of the ability of the adenovirus to escape the endolysosomal pathway and increase DNA delivery to the nucleus. DNA–ligand complexes have been shown to be efficient in airway epithelia in vitro (23). Ferkol and colleagues have also shown that molecular conjugates that use a Fab fragment of IgG against human secretory component can deliver reporter genes to the airway epithelia in vitro and also in vivo following intravenous administration (42,43). These complexes appear to preferentially target the larger cartilaginous airways where expression of the target polymeric IgA receptors predominates.

Recently, DNA polylysine complexes have been linked to other ligands such as the serpin enzyme complex receptor in an attempt to increase airway epithelial gene transfer efficiency (see strategies section below) (196). The poly-L-lysine polymer may also be modified to enhance gene transfer efficiency. Kollen et al. (93) have recently demonstrated that lactosylation of poly-L-lysine can significantly enhance gene transfer to cultured CF cells in vitro.

IN VIVO METHODS OF ADMINISTRATION

Several methods have been devised for delivery of gene transfer vectors to the lung. These include ex vivo delivery, intravenous delivery, direct instillation, and aerosol administration.

Ex Vivo Delivery

The complex geometry of the lung has generally made ex vivo strategies for lung gene transfer untenable. However, an ex vivo strategy—modification of cells in vitro with subsequent injection in vivo—for therapy of pulmonary hypertension has been proposed (16). Campbell et al. genetically modified cultured pulmonary artery smooth muscle cells by transfection with a plasmid containing the human eNOS cDNA using SuperFect™ (Qiagen, Valencia, CA, USA) transfection reagent (16). The modified cells were then injected into the superior vena cava of rats where they migrated into the pulmonary circulation and demonstrated attenuation of monocrotaline-induced pulmonary hypertension. Surprisingly, these smooth muscle cells were localized within the pulmonary arterioles for several days. Thus, ex vivo strategies may be applicable to therapy of pulmonary hypertension.

Intravenous Delivery

Intravenous methods of delivery have been extensively used for delivery of nonviral vectors to the lung. The technique is attractive because of the ease of administration. The species most commonly employed for lung delivery has been the mouse. In this technique, tails of anesthetized mice are warmed in 55° to 60°C water to aid in visualization of the tail vein. Lipoplexes containing plasmid DNA (50–100 μg) in a volume of up to 400 μL are then injected into the tail vein with a 30-gauge needle. This method of delivery primarily leads to uptake of vector into alveolar type II cells and lung endothelial cells (63,101). Airway delivery, as might be required for cystic fibrosis, is limited.

The major disadvantage is potential systemic spread of vector leading to ectopic expression of transferred DNA.

Direct Lung Instillation

Direct instillation into nose and trachea has been the most common method of lung delivery in animals and in human clinical trials. Vector can be delivered by direct instillation onto the nasal epithelium or by intratracheal administration. Nasal delivery in rodents can target both the nasal and lower airways, whereas intratracheal adminstration targets perhaps smaller airways and the alveolar region.

Intranasal Instillation

Two principal methods of nasal delivery have been reported in the literature. One method administers vector drop by drop to the nostrils of anesthetized mice with a micropipet to permit nasal inhalation of vector (123). Up to 100 µL of vector can be delivered by this method (66,193), although smaller volumes (10–30 µL) are routinely used. Alternatively, vector (20–100 µL) can be perfused directly onto the nasal epithelium via a small polyethylene catheter inserted approximately 5 mm into the nostrils of anesthetized mice at a rate of 1.5 µL/minute with a microperfusion pump (193). The catheter is slowly pulled under heat to decrease the diameter of the tip to 20 to 50 µm prior to use. Slow perfusion onto the epithelium increases the residence time of vector in the nasal cavity, which has also increased gene transfer efficiency to the nasal epithelium (193). For human studies, nasal instillation has been accomplished by catheter infusion of vector (0.2–2 mL) under direct vision onto the inferior nasal turbinate (18,57,91,132,187). In some cases, vasoconstrictor agents and topical anesthetics are applied prior to vector instillation (187).

Tracheal Administration

Vector can also be delivered to the tracheas of anesthetized mice using the double tracheostomy technique (77,78,126). Tracheas of anesthetized mice are surgically exposed through a 0.5- to 1.0-cm skin incision and a distal tracheotomy performed near the carina for ventilation (breathing) and a proximal tracheotomy performed at the first cartilaginous tracheal ring for vector instillation. The region between the two tracheostomies is filled with vector for a defined period of time until subsequent removal by suctioning. Limitation of instilled vector to the region between the two tracheostomies by respiration through the distal tracheostomy can be confirmed by visualization through a dissecting microscope. Intratracheal instillations of vector or vehicle can be performed for up to 2 hours, with a mortality rate of approximately 10%. A small animal ventilator is required for longer periods of vector instillation. Closure of the skin alone is sufficient since a tracheostomy tube is not used. However, if a tracheostomy tube is used, the trachea should be repaired prior to skin closure.

Intratracheal/Bronchial Instillation

Two techniques have been employed in rodents: *(i)* blind intubation in which the tongue of an anesthetized mouse is pulled forward, a small animal gavage or intubation needle is passed through the vocal cords into the trachea, and vector (50–100 µL for mice and 300 µL for rats) with air is injected (62), and *(ii)* passage of a 22- to 30-gauge needle into the surgically exposed tracheas of supine mice lying at a 30° to 45° angle followed by injection of vector and air into the trachea (63,170). The former technique is better for delivery to specific regions of lung (left vs. right lung), whereas the latter technique induces less injury to the trachea. For studies in rabbits, nonhuman primates, and humans, fiberoptic bronchoscopic instillation of vector to a specific lobe is often performed (2,8,22,48,198).

Aerosol Administration

Aerosol administration is a common method of vector delivery in nonhuman primate studies, but has also been used in several clinical CF gene transfer safety and efficacy studies. For nasal administration, crude spray devices have been employed to aerosolize vector onto the nasal epithelium including atomizers (18) and spray devices metered to deliver prescribed volumes (100 µL) per actuation (116). Recently, 4.5-mL lipoplex was delivered by a breath-activated nebulizer (Pari LC Jet Plus™; Pari Respiratory Equipment, Midlothian, VA, USA) to the nasal epithelium (2).

Aerosol delivery of gene transfer vectors to the lower airways and alveoli is more complex. A variety of factors must be entertained before undertaking such studies, including *(i)* the type of nebulizer (e.g., compressed air vs. jet nebulizer); *(ii)* the size of the respirable aerosol particles, with particles in humans greater than 1 to 2 µm in diameter targeting the airways and smaller particles targeting the alveolar region; *(iii)* the effect of nebulization on vector infectivity or stability; *(iv)* the depth and frequency of inhalation, which may affect vector deposition in the lung; and *(v)* dilution of gene transfer vectors in the fluid lining the airways and alveoli. Breath-activated jet nebulizers (OPTINEB™ [Air Liquide, Paris, France] and Pari LC Jet Plus) (2,9,109) permit intermittent aerosolization of vector during inspiration with decreased escape of vector into the environment and up to 80% of respirable aerosol particles may have diameters greater than or equal to 2 μm (9). Vector volumes ranging from 1.6 to 20 mL have been nebulized from the reservoir of these jet nebulizers.

BARRIERS TO AIRWAY GENE TRANSFER

The lung has barriers to gene transfer at virtually all levels, including nonspecific barriers and vector-specific barriers. Nonspecific barriers include airway mucus, glycoconjugates, and the inflammatory milieu, while vector-specific barriers include factors affecting binding and entry, e.g., receptor localization and endocytic capacity, nuclear translocation, and factors limiting transgene expression post nuclear entry. These barriers combine to make in vivo gene transfer to human airways inefficient.

Nonspecific Barriers to Gene Transfer

In cystic fibrosis, the manifestation of mutant CFTR as defective ion transport leads to abnormal secretions, ineffective mucociliary clearance, bacterial proliferation with multiresistant organisms, bronchiectasis, and ultimately death (24). Expression of mutant CFTR in airway epithelia is also associated with an inflammatory response characterized by a massive influx of neutrophils, which act as a source of oxidants and proteolytic enzymes promoting tissue injury and bronchiectasis. This influx of neutrophils increases the load of DNA and actin in the airway secretions of the CF lung, leading to a markedly increased sputum viscosity with markedly elevated levels of proinflammatory cytokines including TNFα, IL-1β, IL-6, and IL-8 that perpetuate the inflammatory response (12). This inflammation with abnormal secretions and *Pseudomonas* colonization is established early in life , often in the first year (87,88,115). Because current therapies do not eradicate the chronic pulmonary infection and inflammation of the CF lung, luminal gene transfer approaches targeting the superficial airway epithelium will likely have to overcome the inflammatory response to gain vector access to the epithelium.

Stern et al. investigated the effect of fresh sputum obtained from CF patients on gene transfer to primary airway cells and to CF cell lines in vitro (151). A dose-dependent inhibition of gene transfer to COS-7, 16HBE14o⁻, and 2-CFSMEO cells mediated by the cationic liposome DC-Chol/DOPE and an Ad-*lacZ* vector was detected in the presence of UV light-sterilized CF sputum. The effects of CF sputum on liposomal gene transfer were partially reversible using recombinant DNase (rDNase) pretreatment and completely reversible when rDNase pretreatment preceded Ad gene transfer. However, pretreatment with other mucolytic agents including nacystelyn, lysine, *N*-acetylysteine, and rAlginase failed to increase liposomal or adenoviral gene transfer efficiency. This effect could be simulated by application of genomic DNA to cultures prior to transduction with DC-Chol/DOPE or an Ad-*lacZ* vector. These data suggest that noninfectious sputum components can inhibit gene transfer mediated by cationic liposomes and adenoviruses, and that excessive DNA is a major contributor to this inhibitory response.

The effect of bronchopulmonary inflammation induced by *Pseudomonas aeruginosa* on Ad-mediated gene transfer to airway epithelial cells in vivo has also been investigated (160). An animal model of chronic bronchopulmonary infection, in which mice inoculated with *Pseudomonas aeruginosa*-laden agarose beads, develop bronchitis, bronchopneumonia, bronchiectasis, mucus plugging, and alveolar exudate with acute and chronic inflammatory cells was employed. These studies demonstrated a greater than 2-fold reduction in gene transfer efficiency mediated by nasal instillation of an Ad-*lacZ* vector in mice, in which the *Pseudomonas*-laden beads were instilled, compared to mice that had sterile beads or Ad vector alone instilled. Similarly, Ad-*lacZ* gene transfer to nasal airways of *Pseudomonas* (PAO1 strain)-infected mice reduced gene transfer efficiency 10-fold, relative to noninfected nasal airways (123). Thus, the inflammatory milieu induced by *Pseudomonas* is a formidable barrier to transduction.

Components of the innate immune system, including alveolar macrophages and airway surface fluid, may also serve as barriers. Worgall et al. (177) used Ad-specific probes and Southern analysis to demonstrate a 70% loss of Ad-*lacZ* genomes within 24 hours in both immunocompetent and immunodeficient mice 24 hours following transtracheal administration of an Ad-*lacZ* vector. Elimination of macrophages by administration of liposomes containing dichloromethylene biphosphonate with subsequent administration of the Ad-*lacZ* vector resulted in a significant (100%) increase in lung DNA and subsequent β-galactosidase expression. Unlike the persistence of Ad vector DNA detected in epithelial cell lines, in vitro studies in cultured human alveolar macrophages demonstrated rapid loss of Ad vector genomes. Therefore, alveolar macrophages may play a significant role in reducing the efficiency of Ad gene transfer in the lung parenchyma.

Alveolar macrophages have also been shown to inhibit retrovirus-mediated gene transfer to airway epithelia (108). Transduction of human airway epithelial cells by an amphotropic enveloped retroviral vector was inhibited approximately 40% by alveolar macrophages and by more than 60% by lipopolysaccharide (LPS)-activated alveolar macrophages (108). Incubation of macrophages with dexamethasone (1 μM) partially reversed this inhibition of retroviral transduction. Rapid uptake of labeled vector into vesicles of macrophages was associated with loss of DNA within 24 hours consistent with rapid degradation, rather than rapid transduction, of alveolar macrophages. These data suggest that macrophages can play a significant role in inhibiting in vivo gene transfer to lung epithelia.

Components of the glycocalyx and the airway surface liquid may also have potential inhibitory effects on airway gene transfer. Arcasoy and colleagues (5) demonstrated that overexpression of the mucin MUC1 reduced Ad-mediated gene transfer to MDCK (Madin Darby canine kidney) and bronchial epithelial cells that could be overcome by removal of sialic acid residues from the apical surface by neuraminidase pretreatment. Accordingly, molecules shed from the glycocalyx into the airway surface liquid might also inhibit gene transfer. However, McCray and coworkers (108) demonstrated that airway surface liquid from well-differentiated human epithelial cell cultures harvested in a small volume of distilled water failed to inhibit retroviral transduction to naïve airway cells in vitro. Johnson and coworkers (77) similarly reported that freshly isolated murine bronchoalveolar lavage fluid had no effect on transduction of airway epithelial cells by VSV-G pseudotyped retroviral vectors. The lack of an effect of airway surface fluid obtained by washing the surfaces of well-differentiated airway cell cultures and bronchial xenografts on transduction mediated by AAV vectors have also been reported (31). Although the samples of the airway surface fluid used in these studies were dilute, these data would suggest that insignificant levels of vector inhibitory substances are present within the soluble components of airway surface fluid.

Vector-Specific Barriers to Luminal Airway Gene Transfer

Coincident with the initiation of clinical gene transfer efficacy and safety trials, in vitro and in vivo preclinical studies began to show evidence for inefficient transduction of airway epithelial cells when vector was delivered to the luminal surface. In this section, we review data identifying the barriers to gene transfer for vectors that are either currently used in clinical trials or exhibit potential for use in clinical trials.

Ad Vectors

Despite ongoing clinical trials, preclinical studies rapidly established that luminal gene transfer to airways was inefficient (36,64). Grubb et al. (64) demonstrated that luminal application of an Ad-*lacZ* vector efficiently transduced basal cells, the predominant cell type at the site of mechanical injury in human and mouse tracheal explants, whereas lumen facing columnar cells in uninjured areas were resistant to gene transfer. Pickles et al. (126) confirmed this observation in model systems of well-differentiated rat and human airway epithelia and extended it to human intrapulmonary (bronchial) airways. Parallel experiments in excised human airway specimens demonstrated preferential transduction of undifferentiated regenerating or wound repairing cells by Ad vectors, but not well-differentiated pseudostratified columnar epithelia (33).

Subsequent studies explored the reason for the inefficiency of Ad gene transfer to well-differentiated epithelial cells following luminal application. An early observation was that $\alpha_v\beta_{3/5}$ integrins, which mediate uptake of Ad vectors (60,173), were not expressed on the apical surface of columnar cells, limiting cellular entry and hence, efficient gene transfer (61). Studies of radio-and fluorescent-labeled Ad vectors confirmed that decreased binding to the apical membrane of these polarized airway epithelial cells, as compared to poorly differentiated airway cells, and a low rate of vector internalization, out of proportion to the reduction in binding, were barriers to efficient gene transfer (126,129,163,190). Functional studies demonstrating preferential transduction of polarized well-differentiated airway epithelia, following application of vector to the basolateral membrane as compared to the apical membrane, localized the binding and uptake receptors for Ad to the basolateral membrane (126,129,163,190), which was confirmed by immunofluorescent detection of CAR (129). While the resistance of the epithelium to Ad-mediated gene transfer may be partially overcome by increasing the duration of Ad vector incubation with the epithelium (76,80,131,193), direct measurements of nonspecific (fluid-phase) endocytosis using radiolabeled markers have demonstrated a markedly reduced rate of endocytic uptake in well-differentiated airway epithelia as compared to poorly differentiated airway epithelia (107). Thus, minimal enhancement of gene transfer would be generated by increases in nonspecific binding of Ad vectors to well-differentiated airway epithelia.

The aforementioned studies confirm that vector-specific barriers to Ad-mediated gene transfer are present in the apical membrane of well-differentiated airways, including lack of receptors and integrins leading to decreased uptake and decreased entry of gene transfer vectors. Strategies to improve vector access with luminal application must address these considerations.

AAV Vectors

Evidence is evolving that transduction of well-differentiated airway epithelia by luminal application of AAV vectors may also be inefficient. Summerford et al. (153) have identified a membrane-associated heparan sulfate proteoglycan as a receptor for adeno-associated virus serotype 2 (AAV-2), the most common serotype used in AAV vectors. Fibroblast growth factor-1 (FGF-1) and $\alpha_v\beta_5$ integrins act as coreceptors for AAV (133,152). Heparan sulfate proteoglycans and the coreceptors have also been localized predominantly to the basal surface, but not the apical membrane of well-differentiated airway epithelial cultures (31). Basolateral receptor localization correlates with preferential transduction of these cells when vector is applied to the basolateral membrane relative to the apical membrane.

A post-entry barrier to transduction identified in undifferentiated primary cells is the persistence of AAV genomes as single-stranded episomes, which are inefficiently converted to double-stranded DNA, a requirement for transgene expression (44,45,65). This limitation leads to a delay in the onset of transgene expression, which may be overcome by waiting long enough for maximal transgene expression to occur (approximately 4 weeks) or by the use of DNA damaging agents, topoisomerase inhibitors, and Ad early gene products (1,134, 141). Thus, barriers to AAV transduction include the decreased binding and uptake of vector due to decreased receptor expression on the apical membrane and inefficient single-strand to double-strand conversion of AAV genomes.

Cationic Liposomes

Several undifferentiated cell lines or cell types are resistant to transfection by cationic liposomes (189). Nuclear entry has been identified as the rate-limiting factor for efficient liposome-mediated gene transfer in these cell lines (189), whereas gene transfer to well-differentiated airway epithelial cells is limited by failure of DNA–liposome complexes to enter the cell (20,107). Matsui et al. (107) used rat and human airway cells grown as islands on permeable collagen substrates, in which poorly differentiated cells form on the edges of the islands and polarized (well-differentiated) cells form in the central portions, to investigate the relative efficiency of liposome-mediated gene transfer. Matsui demonstrated loss of phagocytic entry mechanisms, decreased cell surface binding, and decreased uptake in differentiated airway epithelial cells (central cells) as compared to poorly differentiated cells (edge cells) as the reason for inefficient transduction. A subsequent study confirmed these observations by demonstrating decreased amounts of cell-associated lipoplexes in differentiated airway epithelia as compared to poorly differentiated epithelia (41). These investigators also documented enhanced gene transfer in proliferating cells relative to quiescent cells by lipoplexes, raising the possibility of enhanced nuclear transport of DNA during mitosis due to breakdown of the nuclear envelope. The choice of lipid used in these 2 studies did not affect the low efficiency of gene transfer to differentiated epithelia. Thus, barriers to efficient transduction of well-differentiated airway cells by lipoplexes include decreased binding and uptake of lipoplexes and poor nuclear translocation of DNA.

Retroviral Vectors

Retroviral vectors are attractive for CF and other genetic diseases because of their potential for long-term expression as a result of integration into the host cell genome. The lack of cell proliferation in well-differentiated airway epithelia in vivo (99) and low titers have served as major barriers to the use of retroviral vectors derived from Moloney MuLV for in vivo airway gene transfer. The development of lentiviral vectors derived from HIV (113,114,199, 200), EIAV (117), and FIV (167), which can transduce nondividing airway cells, may soon overcome the requirement for cell proliferation in some cell types. Advances in retroviral technology, including production techniques and pseudotyping of vectors to permit concentration of vector stocks, may also soon overcome the limitations of titer (15).

However, other barriers to efficient transduction of well-differentiated airway cells by retroviruses clearly exist. Wang et al. (165) have demonstrated efficient transduction of polarized well-differentiated airway cells stimulated to proliferate with keratinocyte growth factor (KGF) when amphotropic enveloped MuLV vectors were applied to basolateral surface as compared to minimal or no gene transfer when vector was applied to the apical surface. These data are consistent with localization of the amphotropic receptor to the basolateral surface of cultured well-differentiated airway cells. Western blot and molecular data suggest that receptor levels may be low in the absence of KGF (165,166,198).

Similar findings may occur with other enveloped retroviral vectors. Wild-type vesicular stomatitis virus (VSV) has been shown to preferentially infect polarized MDCK cells, a model for polarized airway epithelia, across the basolateral membrane (54). Since the envelope glycoprotein of vesicular stomatitis virus has been used to pseudotype MuLV and lentiviral vectors derived from HIV and EIAV, VSV-G pseudotyped retroviral and lentiviral vectors would also be expected to preferentially transduce polarized MDCK cells from the basolateral surface. This notion has been confirmed in polarized well-differentiated airway cells (78).

Inflammatory and Immune Barriers to Airway Epithelial Gene Transfer

The induction of immune and inflammatory responses by gene transfer vectors may serve as a significant barrier to the use of gene transfer vectors clinically. To date, the responses induced by Ad vectors have been the most pronounced (27,58,83,102,120,150,159,180,183,184). However, humoral immune responses have also been noted with AAV vectors and inflammatory conditions have been detected following delivery of nonviral vectors (8,66,67,71,144).

Adenoviruses

The immunogenic responses induced by Ad vectors are a major concern for gene therapy of lung disease. Ad vectors induce an acute nonspecific mixed cellular inflammatory response and a late specific dose-related, lymphocyte-predominant, cell-mediated immune response in murine, cotton rat, and baboon lungs (27,58,83,102,120,148,150,159,180,183,184). The acute response is nonspecific and likely induced by cytokines, while the later specific immune response is mediated by major histocompatibility complex (MHC) class I-restricted cytotoxic (CD8) T lymphocytes directed against viral gene products and transgene proteins leading to destruction of transduced cells with resulting decreased persistence. Despite the E1-deletion, first generation vectors have been shown to express late gene products, e.g., hexon. Second generation E2a defective vectors reduce late gene expression and inflammation extending the duration of transgene expression in nonhuman primates and in rodents, although the duration of the cellular inflammation is also extended (35,38,59,181).

Second generation E1-deleted Ad2 vectors with deletions of the E4 region, except for ORF6 (see Figure 1B), have also been associated with cellular inflammation. Studies in nonhuman primates examining toxicity with repetitive administration (every 3 weeks for up to 11 doses) of these vectors suggest that histopathologic changes of inflammation are minimal at doses of up to 3×10^9 IU delivered to a single lobe of the lung (150). However, doses of vector greater than 3×10^9 IU generated the expected histologic changes of inflammation. Subsequent studies of Ad2 vectors with wild-type E4 regions, partial deletions of E4 retaining only ORF6, and complete deletions of E4 displayed a complex relationship between promoters and the persistence of gene transfer. Sustained expression of the *lacZ* transgene occurred when driven by a cytomegalovirus (CMV) promoter in nude mice in the presence of wild-type E4 *cis* elements, whereas transgene expression from vectors with deletions of E4 did not persist (6,82). Surprisingly, other promoters did not mediate persistent *lacZ* expression.

High-capacity vectors that have had all of the viral genes deleted (ΔrAd) have not been extensively evaluated in the lung, but studies in the liver suggest an improved safety profile (100,112). Complete removal of helper virus (E1-deleted virus used to produce ΔrAd) from ΔrAd vector is a current limitation to the widespread use of completely deleted vectors.

Humoral responses to Ad vectors include the development of mucosal and neutralizing antibodies (83,104,106,120,142,160,182,184). This immune response is secondary to a helper (CD4+) T lymphocyte response directed against the adenoviral capsid proteins. The major limitation of these antibodies is the prevention of Ad infection on subsequent readministration of vector.

Other potential limitations include neurogenic inflammation and dose-dependent increases in apoptosis or programmed cell death (130,157). Intra-airway administration of E1, E3-deleted Ad vectors induced a dose-dependent potentiation in capsaicin stimulated vascular permeability in rat airways consistent with neurogenic inflammation (130). This dose-dependent effect was reduced by UV-psoralen inactivation of the Ad vector, consistent with inhibition of viral gene expression, and by administration of a selective substance P (NK1) receptor antagonist. Dose-dependent induction of apoptosis and cell cycle alterations occurring after Ad infection of airway cells may be harmful to the reparative responses in the lung (157).

AAV

Cell-mediated immune responses have not been reported in ongoing preclinical and clinical studies in the lungs of animals dosed with AAV vectors (162). However, contradictory results have been published regarding the generation of a humoral immune response. Halbert et al. (66) demonstrated transduction of approximately 5% to 6% of rabbit airway cells by an AAV alkaline phosphatase and *lacZ* vectors following balloon treatment of rabbit airway. However, AAV failed to transduce balloon-treated right lower lobe (RLL) airway epithelia of rabbits pretreated with AAV to the left lower lobe (LLL) 14 days previously. This failure to transduce rabbit airways upon readministration of vector correlated with an increase in neutralizing antibody that could be overcome by transient immunosuppression. Beck et al. (8) also reported increased levels of neutralizing antibody to AAV in rabbits following nasal and bronchoscopic administration. However, no inhibition of AAV gene transfer was detected in their study despite the increased neutralizing antibody. Reasons for the differences between these studies are unknown, but may include differences in delivery technique with greater injury in the balloon catheter model leading to greater antigen exposure and wild-type Ad contamination of AAV vector stocks. Further studies of AAV produced from helper virus-free systems are required to fully answer the role of neutralizing antibody-mediated inhibition of gene transfer following repetitive administration of AAV vector.

Cationic Liposomes

Liposomes have generally been associated with a good safety profile in vivo. However, the low-toxicity profile (17,96), which has generally made liposome-mediated gene transfer attractive, may not be a feature of some newer lipids (97,144). GL-67 has been shown to induce an acute dose-dependent neutrophil predominant inflammatory response following intratracheal administration to murine lung, which tends to be less severe following aerosol administration (97,144). In a recent CF clinical study, bronchoscopic, but not nasal administration of CFTR/GL-67 lipoplexes induced an influenza-like syndrome characterized by fever, myalgias, and headache with negative chest X-rays and blood cultures (2). The etiology of this inflammatory response is unclear, although unmethylated plasmid DNA and the lipid have both been entertained. However, the inflammation per se does not demonstrate a clear effect on either efficiency or persistence of gene transfer.

In summary, a variety of barriers to efficient transduction of airways in vivo by luminal application of vector exist. Decreased binding and uptake of vector due to lack of apical membrane receptors and internalization pathways appears to be a common barrier to all vectors. Other barriers such as second strand synthesis and nuclear entry are more vector specific. Thus, strategies that permit better apical membrane entry or that allow access of the vector to the basolateral surface are likely to be generally useful. Circumventing the immune response will also be important for transient expression vectors.

STRATEGIES TO OVERCOME APICAL MEMBRANE BARRIERS TO EFFICIENT TRANSDUCTION

Since the apical membrane of well-differentiated cells is a major barrier to efficient transduction by all the current gene transfer vectors, strategies to overcome apical membrane barriers are crucial. Although intravenous approaches have been considered (42,43), the multiple barriers that must be crossed (e.g., endothelium, endothelial basement membrane, interstitium, and epithelial basement membrane) have made luminal delivery of vectors attractive. Two strategies for improving gene transfer efficiency using luminal application of vector have been proposed. One strategy focuses on modification of the host by increasing the paracellular permeability of airway epithelia to allow access of vector to the receptors on the basolateral membrane of the airway cells, and the other strategy modifies vectors to target receptors expressed on the apical membrane of airways in vivo that have the capacity to internalize.

Host Modification

A number of agents have been used in the literature to modify paracellular permeability (3,103). These agents can generally be grouped into agents that have relatively nonspecific effects and agents that target specific components of the intercellular junction. In vitro and in vivo studies have recently begun to appear in the literature that establish the feasibility of such approaches. Inhalation of the oxidant gas sulfur dioxide promotes denuding of the surface epithelium in a dose-dependent manner, while also increasing paracellular permeability in less severely injured regions (72,77,78). Johnson and colleagues (77,78) demonstrated that this oxidant model could be used to stimulate epithelial cell proliferation and enable relatively efficient gene transfer to the airways of mice using a VSV-G pseudotyped retroviral vector. In a subsequent study, Parsons and colleagues (123) demonstrated that low doses of the surface active detergent or enhancer polidocanol, increased airway permeability (measured by lanthanum permeation into the intercellular junctions) of polidocanol-treated murine airways, but not that of control animals without inducing frank morphologic injury. Pretreatment of nasal airways with this surface active agent enhanced gene transfer mediated by an Ad-*lacZ* vector and facilitated partial correction of the Cl⁻ transport defect in the nasal epithelium of CF mice following a single dose of vector. The single dose of Ad-CFTR vector used following pretreatment with the surface agent polidocanol generated the same degree of CFTR correction previously reported by Grubb (64), in which four doses of an Ad vector were required to generate a 40% to 50% correction of Cl⁻ transport.

The adherens junction of the intercellular junctional complex has become a target for strategies directed at transient permeabilization of airways to enhance gene transfer. Wang et al. (165) demonstrated enhanced retroviral gene transfer mediated by an amphotropic enveloped vector applied to the apical membrane of well-differentiated airway epithelia following pretreatment with the calcium chelator EGTA and hypotonic solutions. Increasing paracellular permeability using the EGTA/hypotonic solution treatment enabled the investigators to correct the Cl⁻ transport defect in well-differentiated human CF airway epithelial cell cultures stimulated to proliferate with KGF when vector was applied to the luminal surface. Subsequently, these investigators used EGTA and hypotonic solution to enhance FIV-based lentiviral gene transfer to rabbit airways in vivo (167). A 7- to 10-fold increase in AAV-mediated transduction of well-differentiated primary airway cell cultures using an AAV green fluorescent protein (GFP) vector and transient permeabilization with EGTA/hypotonic solution has also been reported (31). These studies establish the feasibility of using the transient modification of the host by permeabilization of the paracellular path to enhance airway gene transfer in vivo mediated by vectors whose receptors are located on the basolateral membrane.

Vector Modification

Modifications to specific gene transfer vectors is another approach to enhancing gene transfer efficiency. This strategy focuses on targeting the vector to receptors that are specifically expressed on the apical membrane of well-differentiated human airway epithelial cells. The concept of targeting gene transfer vectors to alternative (non-wild-type) receptors is well established in the literature for adenoviral, retroviral, and nonviral vectors (21,26,28,76,85,89,145-

147,169,172,174,179,186). Three general strategies have been used for Ad vectors: *(i)* genetic engineering of peptide ligand sequences or single chain antibody fragments (scFv) into the fiber knob domain of Ad (26,174); *(ii)* the use of scFv-fusion proteins with specificities for an Ad epitope while bearing a ligand binding domain for a specific cell surface receptor, the adenobody approach (169); and *(iii)* the use of bispecific antibodies composed of two antibodies—one directed against the Ad vector and the other against the specific cell surface receptor—that have been cross-linked (172). Similar strategies have been employed with retroviral vectors with the creation of chimeric envelopes bearing a ligand or single chain antibody specific for a cell surface receptor in the N-terminal domain of envelope protein (21,85,145,179), or as has recently been reported, fused directly to transmembrane domain of envelope with the use of a spacer (76). Incorporation of ligand molecules into the lipid bilayers of liposomes is well described, and biotinylation of ligands for use with streptavidin-labeled DNA polymers has also been reported (146,147).

While each of these methods has been successful at increasing binding, successful vector-specific transduction has been more difficult. Genetic incorporation of ligands is limited by size of the peptide sequence and the ability to incorporate the chimeras into viral particles, while bispecific antibodies and adenobodies must fold properly to bind specifically and enter with the target receptor (21,121,145). Furthermore, a membrane or endosomolytic fusion event must occur for the vector to gain access to the cytoplasm and henceforth, the nucleus. Nevertheless, the aforementioned approaches have been successfully used in vitro to target a variety of receptors including α_v integrins (173), the epidermal growth factor receptor (169), stem cell factor receptor (147,179), and T-cell receptors such as CD3 (172).

Since most growth receptors are expressed on the basolateral membrane of well-differentiated cells, major concerns are which receptors are expressed on the apical membrane and whether the levels of expression are sufficient to promote successful gene transfer. $P2Y_2$-R, a seven transmembrane purinoceptor that normally mediates acute airway epithelial cell responses to the luminal environment, the serpin enzyme complex receptor (SEC-R), and the urokinase plasminogen activator receptor (uPA-R) have each been proposed as targets on the apical membrane (29,73,105,196). Studies in polarized cells have documented that bispecific monoclonal antibody Ad vector complexes directed toward epitope-tagged external domains of $P2Y_2$-R can promote enhanced gene transfer efficiency (127,128) and targeted transduction of wild-type $P2Y_2$-R expressing cell lines mediated by vector conjugates composed of Ad complexed to chemically modified ligands such as biotin-UTP, has also been demonstrated in preliminary studies (95). A uPA-polyethylene glycol (PEG)-coated adenoviral (Ad) vector has also been shown to enhance Ad gene transfer to polarized airway epithelia (29,119). Recently, the SEC-R has been used to target DNA–polylysine–ligand complexes to airways in vitro and in vivo (196). Thus, targeting of vectors to alternative apical membrane receptors may be applicable to a variety of vectors.

Circumventing the Immune Response to Enhance Gene Transfer

This strategy is most applicable to cell-mediated and humoral immune responses to Ad vectors and potentially to humoral responses mediated by AAV (67,86,102). Two major approaches have been developed to blunt the cell-mediated responses have been tested. The first strategy is based on the identification of specific genes that initiate or amplify the cell-mediated immune responses, permitting subsequent vector reengineering to delete their expression. Examples of this approach include the removal of specific viral gene products (e.g., E2A, E4) and complete deletions of all genes in the vectors that express viral proteins (e.g., high-capacity vectors) (6,35,38,45,59,68,69, 92,100,112,122,181). Alternatively, vector genes that have evolved to subvert antiviral mechanisms in host cells may be reinserted into completely or partially deleted Ad vectors to modulate host–cell defense responses (6,14). For example, the 19-kDa glycoprotein encoded in the Ad E3 region has been shown to reduce cytotoxicity to virus-infected host cells by inhibiting MHC class I transport to the cell surface (14). An adenoviral vector in which high-level E3 expression was ensured by promoter selection has been demonstrated to improve transgene persistence in an animal model, and some investigators have demonstrated increased persistence transgene expression by retention of the E4 region (6,14,84).

Two strategies have also been proposed to overcome the humoral immune response that leads to inhibition of repetitive viral vector administration: *(i)* serotype switching (104,106), and *(ii)* transient immunosuppression (67,81, 176). With regard to serotype switching, alternate dosing with Ad vectors from different subgroups and even different serotypes has been proposed as an approach to circumvent the anti-Ad humoral response limiting gene transfer with repetitive administration (104,106). Intratracheal administration of wild-type Ad5 (subgroup C), but not wild-type Ad4 (subgroup E) or wild-type Ad30 (subgroup D), has been reported to prevent subsequent gene transfer mediated by intratracheal administration of an Ad5-*CAT* or Ad5-*lacZ* vector (subgroup C) seven days later (106). Alternatively, transient immunosuppression may prevent inhibition of Ad gene transfer by neutralizing antibody production upon readminsitration of vector. Yang et al. (182) have reported that intratracheal and/or intraperitoneal delivery of IL-12 and interferon gamma (IFNγ) markedly reduced (60-fold) production of vector-specific neutralizing and IgA antibodies in bronchoalveolar lavage (BAL) fluid of mice infected with an Ad5 vector. Furthermore, co-administration of IL-12 and IFNγ with an Ad-*lacZ* vector permitted successful gene transfer 28 days after previous intratracheal dosing with an Ad-*ALP* (alkaline phosphatase) vector. Blockade of T-cell co-stimulatory pathways with antibodies against CD40 ligand and CTLA4, and the use of pharmacological agents such as corticosteroids and cyclophosphamide at the time of vector administration, have also been shown to inhibit the development of neutralizing antibodies to Ad vectors (67,81,176). Blockade of co-stimulatory pathways at the time of AAV administration has also been shown to inhibit neutralizing antibody to AAV and permit successful AAV gene transfer following readministration (67). Recently, coating of Ad vector capsids by covalent linkage of polyethylene glycol (PEG) and related compounds has been suggested as one approach by which Ad vectors might escape antibody neutralization in mice with high neutralizing antibody titers (119). This approach, although promising, is currently limited by the deleterious effects of PEG and related compounds on vector titer and aggregation (119).

Other Approaches

Polycations have also been used to enhance airway gene transfer. A variety of polycations, including poly-L-lysine, diethylaminoethyl (DEAE)-dextran, polybrene, and protamine sulfate, have been shown to enhance Ad-mediated gene transfer to airway epithelia in vitro and in vivo presumably by decreasing adverse cell charge interactions (4,84). Polycations have also been shown to increase Ad vector binding and uptake in cell lines and undifferentiated primary airway epithelial cell cultures (4). However, a role for changes in epithelial permeability as a mechanism of gene transfer enhancement has not been excluded. Further safety studies of polycations are required, although toxicity appears to be limited with current concentrations of polycations used. Both co-administration of equal volumes of polycation and Ad vector preincubated for 30 minutes prior to delivery, as well as pretreatment of the epithelium with polycation prior to vector application, appear to enhance gene transfer.

Similarly, calcium phosphate coprecipitation has been used to create aggregates of Ad that lead to enhanced non-CAR-dependent binding and uptake of Ad vectors in National Institutes of Health (NIH) 3T3 cells (164). These coprecipitates also increase Ad and AAV-mediated gene transfer to airway epithelia in vitro and in vivo (40,164). Whether this mechanism is operative in vivo, where the rate of endocytosis across the apical membrane by nonspecific mechanisms is low, has not been delineated. Possible effects on cellular and paracellular permeability have not been investigated.

SUMMARY

A number of methods are now available to deliver genes to airways and alveolar regions of the lung. Significant advances have also been made in the design and production of vectors for lung gene transfer. However, the lung remains a formidable target for gene transfer approaches because of its complex anatomy, large surface area, protective immune functions, and the presence of specific barriers that limit entry of foreign antigens. Therefore, a more thorough understanding of the lung biology relevant to gene transfer is required, if present and future gene transfer approaches for lung disease are to be successful.

ACKNOWLEDGMENTS

The author would like to thank Ms. Miriam Kelly and Ms. Carolyn Coyne for their assistance in the preparation of this manuscript. This work was supported in part by Grant No. HL58342 from the National Heart Lung and Blood Institute.

REFERENCES

1. Alexander, I.E., D.W. Russell, and A.D. Miller. 1994. DNA-damaging agents greatly increase the transduction. J. Virol. *68*:8282-8287.
2. Alton, E.W.F.W., M. Stern, R. Farley, A. Jaffe, S.L. Chadwick, J. Phillips, J. Davies, S.N. Smith, J. Browning, M.G. Davies, et al. 1999. Cationic lipid-mediated CFTR gene transfer to the lungs and nose of patients with cystic fibrosis: a double-blind placebo-controlled trial. Lancet *353*:947-954.
3. Anderson, J.M. and C.M. Van Itallie. 1995. Tight junctions and the molecular basis for regulation of paracellular permeability. Am. J. Physiol. *269*:G467-G475.
4. Arcasoy, S.M., J.D. Latoche, M. Gondor, B.R. Pitt, and J.M. Pilewski. 1997. Polycations increase the efficiency of adenovirus-mediated gene transfer to epithelial and endothelial cells *in vitro*. Gene Ther. *4*:32-38.
5. Arcasoy, S.M., J. Latoche, M. Gondor, S.C. Watkins, R.A. Henderson, R. Hughey, O.J. Finn, and J.M. Pilewski. 1997. MUC1 and other sialoglycoconjugates inhibit adenovirus-mediated gene transfer to epithelial cells. Am. J. Respir. Cell Mol. Biol. *17*:422-435.
6. Armentano, D., J. Zabner, C. Sacks, C.C. Sookdeo, M.P. Smith, J.A. St.George, S.C. Wadsworth, A.E. Smith, and R.J. Gregory. 1997. Effect of the E4 region on the persistence of transgene expression from adenovirus vectors. J. Virol. *71*:2408-2416.
7. Barst, R.J., L.J. Rubin, W.A. Long, M.D. McGoon, S. Rich, D.B. Badesch, B.M. Groves, V.F. Tapson, R.C. Bourge, B.H. Brundage, et al. 1996. A comparison of continuous intravenous epoprostenol (prostacyclin) with conventional therapy for primary pulmonary hypertension. N. Engl. J. Med. *334*:296-301.
8. Beck, S.E., L.A. Jones, K. Chesnut, S.M. Walsh, T.C. Reynolds, B.J. Carter, F.B. Askin, T.R. Flotte, and W.B. Guggino. 1999. Repeated delivery of adeno-associated viurs vectors to the rabbit airway. J. Virol. *73*:9446-9455.
9. Bellon, G., L. Michel-Calemard, D. Thouvenot, V. Jagneaus, F. Poitevin, C. Malcus, N. Accart, M.P. Layani, M. Aymard, H. Bernon, et al. 1997. Aerosol administration of a recombinant adenovirus expressing CFTR to cystic fibrosis patients: a phase I clinical trial. Hum. Gene Ther. *8*:15-25.
10. Bergelson, J.M., J.A. Cunningham, G. Droguett, E.A. Kurt-Jones, A. Krithivas, J.S. Hong, M.S. Horwitz, R.L. Crowell, and R.W. Finberg. 1997. Isolation of a common receptor for coxsackie B viruses and adenoviruses 2 and 5. Science *275*:1320-1323.
11. Berkner, K. 1998. Development of adenovirus vectors for the expression of heterologous genes.

BioTechniques *6*:616-629.

12. **Bonfil, T.L.** 1995. Inflammatory cytokines in cystic fibrosis lungs. Am. J. Respir. Crit. Care Med. *152*:2111-21118.

13. **Brasfield, D., G. Hicks, S.J. Soong, J. Peters, and R.E. Tiller.** 1980. Evaluation of a scoring system of the chest radiograph in cystic fibrosis: a collaborative study. Am. J. Roentgen. *134*:1195-1198.

14. **Bruder, J.T., T. Jie, D.L. McVey, and I. Kovesdi.** 1997. Expression of gp19K increases the persistence of transgene expression from an adenovirus vector in the mouse lung and liver. J. Virol. *71*:7623-7628.

15. **Burns, J.C., T. Friedmann, W. Driever, M. Burrascano, and J.-K. Yee.** 1993. Vesicular stomatitis virus G glycoprotein pseudotyped retroviral vectors: concentration to very high titer and efficient gene transfer into mammalian and nonmammalian cells. Proc. Natl. Acad. Sci. USA *90*:8033-8037.

16. **Campbell, A.I.M., M.A. Kuliszewski, and D.J. Stewart.** 1999. Cell based gene transfer to the pulmonary vasculature: endothelial nitric oxide synthase overexpression inhibits monocrotaline induced pulmonary hypertension. Am. J. Respir. Cell Mol. Biol. *21*:567-575.

17. **Canonico, A.E., J.D. Plitman, J.T. Conary, and K.L. Brigham.** 1994. No lung toxicity after repeated aerosol or intravenous delivery of plasmid-cationic liposome complexes. J. Appl. Physiol. *77*:415-419.

18. **Caplen, N.J., E.W.F.W. Alton, P.G. Middleton, J.R. Dorin, B.J. Stevenson, X. Gao, S.R. Durham, P.K. Jeffery, M.E. Hodson, C. Coutelle, et al.** 1995. Liposome-mediated CFTR gene transfer to the nasal epithelium of patients with cystic fibrosis. Nat. Med. *1*:39-46.

19. **Carter, B.J.** 1990. Parvoviruses as vectors, p. 247-284. *In* P.L. Tijssen (Ed.), Handbook of Parvoviruses, Vol 2. CRC Press, Boca Raton.

20. **Chu, Q., J.D. Tousignant, S. Fang, C. Jiang, L.H. Chen, S.H. Cheng, R.K. Scheule, and S.J. Eastman.** 1999. Binding and uptake of cationic lipid: pDNA complexes by polarized airway epithelial cells. Hum. Gene Ther. *10*:25-36.

21. **Cosset, F.L., F. Morling, Y. Takeuchi, R.A. Weiss, M.K.L. Collins, and S.J. Russel.** 1995. Retroviral retargeting by envelopes expressing an N-terminal binding domain. J. Virol. *69*:6314-6322.

22. **Crystal, R.G., N.G. McElvaney, M.A. Rosenfeld, C.-S. Chu, A. Mastrangeli, J.G. Hay, S.L. Brody, H.A. Jaffe, N.T. Eissa, and C. Danel.** 1994. Administration of an adenovirus containing the human CFTR cDNA to the respiratory tract of individuals with cystic fibrosis. Nat. Genet. *8*:42-51.

23. **Curiel, D.T., S. Agarwal, M.U. Romer, E. Wagner, M. Cotton, M.L. Birnstiel, and R.C. Boucher.** 1992. Gene transfer to respiratory epithelial cells via the receptor-mediated endocytosis pathway. Am. J. Respir. Cell Mol. Biol. *6*:247-252.

24. **Davis, P.B.** 1999. Clinical pathophysiology and manifestations of lung disease, p. 45-67. *In* J.R. Yankaskas and M.R. Knowles (Eds.), Cystic Fibrosis in Adults. Lippincott-Raven, Philadelphia.

25. **Delviks, K.A., W.-S. Hu, and V.K. Pathak.** 1997. Ψ-vectors: murine leukemia virus-based self-inactivating and self-activating retroviral vectors. J. Virol. *71*:6218-6224.

26. **Dmitriev, I., V. Krasnykh, C.R. Miller, M. Wang, E. Kashentseva, G. Mikheeva, N. Belousova, and D.T. Curiel.** 1998. An adenovirus vector with genetically modified fibers demonstrates expanded tropism via utilization of a coxsackie virus and adenovirus receptor-independent cell entry mechanism. J. Virol. *72*:9706-9713.

27. **Dong, J.-Y., D. Wang, F.W. Van Ginkel, D.W. Pascual, and R.A. Frizzell.** 1996. Systematic analysis of repeated gene delivery into animal lungs with a recombinant adenovirus vector. Hum. Gene Ther. *7*:319-331.

28. **Douglas, J.T., B.E. Rogers, M.E. Rosenfeld, S.I. Michael, M. Feng, and D.T. Curiel.** 1996. Targeted gene delivery by tropism modified adenoviral vectors. Nat. Biotechnol. *14*:1574-1578.

29. **Drapkin, P.T., J. Zabner, C. O'Riordan, and M.J. Welsh.** 1999. Targeting adenovirus binding to the apical membrane of human airway epithelia [abstract]. Ped. Pulmonol. (Suppl.) *19*:224.

30. **Drumm, M.L., H.A. Pope, W.H. Cliff, J.M. Rommens, S.A. Marvin, L.C. Tsui, F.S. Collins, R.A. Frizzell, and J.M. Wilson.** 1990. Correction of the cystic fibrosis defect *in vitro* by retrovirus-mediated gene transfer. Cell *62*:1227-1223.

31. **Duan, D.-S., Y.-P. Yue, Z. Yan, P.B. McCray, Jr., and J.F. Engelhardt.** 1998. Polarity influences the efficiency of recombinant adeno-associated virus infection in differentiated airway epithelia. Hum. Gene Ther. *9*:2761-2776.

32. **Dubinett, S.M., P.W. Miller, S. Sharma, and R.K. Batra.** 1998. Gene therapy for lung cancer. Hematol. Oncol. Clin. N.A. *12*:569-584.

33. Dupuit, F., J.H.-M. Zahm, D. Pierrot, S. Brezillon, N. Bonnet, J.-L. Imler, A. Pavirani, and E. Puchelle. 1995. Regenerating cells in human airway surface epithelium represent preferential targets for recombinant adenovirus. Hum. Gene Ther. *6*:1185-1193.

34. Egan, M., T. Flotte, S. Afione, R. Solow, P.L. Zeitlin, B.J. Carter, and W.B. Guggino. 1992. Defective regulation of outwardly rectifying Cl⁻ channels by protein kinase A corrected by the insertion of CFTR. Nature *358*:581-584.

35. Engelhardt, J.F., L. Litzky, and J.M. Wilson. 1994. Prolonged transgene expression in cotton rat lung with recombinant adenoviruses defective in E2a. Hum. Gene Ther. *5*:1217-1229.

36. Engelhardt, J.F., R.H. Simon, Y. Yang, M. Zepeda, S.W. Pendleton, B. Doranz, M. Grossman, and J.M. Wilson. 1993. Ad-mediated transfer of the CFTR gene to lung of non-human primates: biological efficacy study. Hum. Gene Ther. *4*:759-769.

37. Engelhardt, J.F., Y. Yang, L.D. Stratford-Perricaudet, E.D. Allen, K. Kozarsky, M. Perricaudet, J.R. Yankaskas, and J.M. Wilson. 1993. Direct gene transfer of human CFTR into human bronchial epithelia of xenografts with E1-deleted adenoviruses. Nat. Genet. *4*:27-34.

38. Engelhardt, J.F., X. Ye, B. Doranze, and J.M. Wilson. 1994. Ablation of E2a in recombinant adenoviruses improves transgene persistence and decreases immune response in mouse liver. Proc. Natl. Acad. Sci. USA *91*:6196-6200.

39. Fagan, K.A., B.W. Fouty, R.C. Tyler, K.G. Morris, Jr., L.K. Hepler, K. Sato, T.D. LeCras, S.H. Abman, H.D. Weinberger, P.L. Huang, et al. 1999. The pulmonary circulation of homozygous or heterozygous eNOS-null mice is hyperresponsive to mild hypoxia. J. Clin. Invest. *103*:291-299.

40. Fasbender, A., J.H. Lee, R.W. Walters, T.O. Moninger, J. Zabner, and M.J. Welsh. 1998. Incorporation of adenovirus in calcium phosphate precipitates enhances gene transfer to airway epithelia *in vitro* and *in vivo*. J. Clin. Invest. *102*:184-193.

41. Fasbender, A., J. Zabner, B.G. Zeiher, and M.J. Welsh. 1997. A low rate of cell proliferation and reduced DNA uptake limit cationic lipid-mediated gene transfer to primary cultures of ciliated human airway epithelia. Gene Ther. *4*:1173-1180.

42. Ferkol, T., C. Kaetzel, and P.B. Davis. 1993. Gene transfer into respiratory epithelial cells by targeting the polymeric immunoglobulin receptor. J. Clin. Invest. *92*:2394-2400.

43. Ferkol, T., J.C. Perales, E. Eckman, C.S. Kaetzel, R.W. Hanson, and P.B. Davis. 1995. Gene transfer into the airway epithelium of animals by targeting the polymeric immunoglobulin receptor. J. Clin. Invest. *95*:493-502.

44. Ferrari, F.K., T. Samulski, T. Shenk, and R.J. Samulski. 1996. Second-strand synthesis is a rate-limiting step for efficient transduction by recombinant adeno-associated virus vectors. J. Virol. *70*:3227-3234.

45. Fisher, K.J., G.P. Gao, M.D. Weitzman, R. DeMatteo, J.F. Burda, and J.M. Wilson. 1996. Transduction with recombinant adeno-associated virus for gene therapy is limited by leading-strand synthesis. J. Virol. *70*:520-532.

46. Fisher, K.J., H. Choi, J. Burda, S.J. Chen, and J.M. Wilson. 1996. Recombinant adenovirus deleted of all viral genes for gene therapy of cystic fibrosis. Virology *217*:11-22.

47. Fitzgerald, D.J., R. Padmanabhan, I. Pastan, and M.C. Willingham. 1983. Adenovirus-induced release of epidermal growth factor and pseudomonas toxin into the cytosol of KB cells during receptor-mediated endocytosis. Cell *32*:607-617.

48. Flotte, T.R., S.A. Afione, C. Conrad, S.A. McGrath, R. Solow, H. Oka, P.L. Zeitlin, W.B. Guggino, and B.J. Carter. 1993. Stable *in vivo* expression of the cystic fibrosis transmembrane conductance regulator with an adeno-associated virus vector. Proc. Natl. Acad. Sci. USA *90*:10613-10617.

49. Flotte, T.R., S.A. Afione, R. Solow, M.L. Drumm, D. Markakis, W.B. Guggino, P.L. Zeitlin, and B.J. Carter. 1993. Expression of the cystic fibrosis transmembrane conductance regulator from a novel adeno-associated virus promoter. J. Biol. Chem. *268*:3781-3790.

50. Flotte, T.R., S.A. Afione, and P.L. Zeitlin. 1994. Adeno-associated virus vector gene expression occurs in nondividing cells in the absence of vector DNA integration. Am. J. Respir. Cell Mol. Biol. *11*:517-521.

51. Flotte, T.R. and B.J. Carter. 1995. Adeno-associated virus vectors for gene therapy. Gene Ther. *2*:357-362.

52. Flotte, T.R., R. Solow, R.A. Owens, S. Afione, P.L. Zeitlin, and B.J. Carter. 1992. Gene expression from adeno-associated virus vectors in airway epithelial cells. Am. J. Respir. Cell Mol. Biol. *7*:349-356.

53. Frederiksen, K.S., A. Petri, N. Abrahamsen, and H.S. Poulsen. 1999. Gene therapy for lung cancer. Lung Cancer *23*:191-207.

54. Fuller, S., C.H. Von Bonsdorff, and K. Simons. 1984. Vesicular stomatitis virus infects and matures only through the basolateral surface of the polarized epithelial cell line, MDCK. Cell *38*:65-77.

55. Gaine, S.P. and L.J. Rubin. 1998. Primary pulmonary hypertension. Lancet *352*:719-725.

56. Geraci, M.W., B. Gao, D.C. Shepherd, M.D. Moore, J.Y. Westcott, K.A. Fagan, L.A. Alger, R.M. Tuder, and N.F. Voelkel. 1999. Pulmonary prostacyclin synthase overexpression in transgenic mice protects against development of hypoxic pulmonary hypertension. J. Clin. Invest. *103*:1509-1515

57. Gill, D.R., K.W. Southern, K.A. Mofford, T. Seddon, L. Huang, F. Sorgi, A. Thomson, L.J. MacVinish, R. Ratcliff, D. Bilton, et al. 1997. A placebo-controlled study of liposome-mediated gene transfer to the nasal epithelium of patients with cystic fibrosis. Gene Ther. *4*:199-209.

58. Ginsberg, H.S., U. Lundholm-Beauchamp, R.L. Horswood, B. Pernis, W.S. Wold, R.M. Chanock, and G.A. Prince. 1989. Role of early region 3 (E3) in pathogenesis of adenovirus disease. Proc. Natl. Acad. Sci. *86*:3823-3837.

59. Goldman, M.J., L.A. Litzky, J.F. Engelhardt, and J.M. Wilson. 1995. Transfer of the CFTR gene to the lung of nonhuman primates with E1-deleted, E2a-defective recombinant adenoviruses: a preclinical toxicology study. Hum. Gene Ther. *6*:839-851.

60. Goldman, M., Q. Su, and J.M. Wilson. 1996. Gradient of RGD-dependent entry of adenoviral vector in nasal and intrapulmonary epithelia: implications for gene therapy of cystic fibrosis. Gene Ther. *3*:811-818.

61. Goldman, M.J. and J.M. Wilson. 1995. Expression of $\alpha_v\beta_5$ integrin is necessary for efficient Ad-mediated gene transfer in the human airway. J. Virol. *69*:5951-5958.

62. Gorman, C.M., M. Aikawa, B. Fox, C. Lapuz, B. Michaud, H. Nguyen, E. Roche, and T. Sawa. 1997. Efficient *in vivo* delivery of DNA to pulmonary cells using the novel lipid EDMPC. Gene Ther. *4*:983-992.

63. Griesenbach, U., A. Chonn, R. Cassady, V. Hannam, C. Ackerley, M. Post, A.K. Tanswell, K. Olek, H. O'Brodovich, and L.-C. Tsui. 1998. Comparison between intratracheal and intravenous administration of liposome-DNA complexes for cystic fibrosis lung gene therapy. Gene Ther. *5*:181-188.

64. Grubb, B.R., R.J. Pickles, H. Ye, J.R. Yankaskas, R.N. Vick, J.F. Engelhardt, J.M. Wilson, L.G. Johnson, and R.C. Boucher. 1994. Inefficient gene transfer by adenovirus vector to cystic fibrosis airway epithelia of mice and humans. Nature *371*:802-806.

65. Halbert, C.L., I.E. Alexander, G.M. Wolgamot, and A.D. Miller. 1995. Adeno-associated virus vectors transduce primary cells much less efficiently than immortalized cells. J. Virol. *69*:1473-1479.

66. Halbert, C.L., T.A. Standaert, M.L. Aitken, I.E. Alexander, D.W. Russell, and A.D. Miller. 1997. Transduction by adeno-associated virus vectors in the rabbit airway: efficiency, persistence, and readministration. J. Virol. *71*:5932-5941.

67. Halbert, C.L., T.A. Standaert, C.B. Wilson, and A.D. Miller. 1998. Successful readministration of adeno-associated virus vectors to the mouse lung requires transient immunosuppression during initial exposure. J. Virol. *72*:9795-9805.

68. Hardy, S., M. Kitamura, T. Harris-Stansil, Y.M. Dai, and M.L. Phipps. 1997. Construction of adenovirus vectors through Cre-lox recombination. J. Virol. *71*:1842-1849.

69. Hartigan-O'Connor, D., A. Amalfitano, and J.S. Chamberlain. 1999. Improved production of gutted adenovirus in cells expressing adenovirus preterminal protein and DNA polymerase. J. Virol. *73*:7835-7841.

70. Hay, J.G., N.G. McElvaney, J. Herena, and R.G. Crystal. 1995. Modification of nasal epithelial potential differences of individuals with cystic fibrosis consequent to local administration of a normal CFTR cDNA adenovirus gene transfer vector. Hum. Gene Ther. *6*:1487-1496.

71. Hernandez, Y.J., J. Wang, W.G. Kearns, S. Loiler, A. Poirier, and T.R. Flotte. 1999. Latent adeno-associated virus infection elicits humoral but not cell-mediated immune responses in a nonhuman primate model. J. Virol. *73*:8549-8558.

72. Hulbert, W.C., S.F. Man, M.K. Rosychuk, G. Braybrook, and J.G. Mehta. 1989. The response phase—the first six hours after acute airway injury by SO_2 inhalation: an *in vivo* and *in vitro* study. Scanning Microsc. *3*:369-378.

73. Hwang, T.H., E.M. Schwiebert, and W.B. Guggino. 1996. Apical and basolateral ATP stimulates tracheal epithelial chloride secretion via multiple purinergic receptors. Am. J. Physiol. *270*:C1611-C1623.

74. Janssens, S.P., K.D. Bloch, Z. Nong, R.D. Gerard, P. Zoldhelyi, and D. Collen. 1996. Adenoviral-mediated transfer of the human endothelial nitric oxide synthase gene reduces acute hypoxic pulmonary vasoconstriction in rats. J. Clin. Invest. *98*:317-324.

75. Jiang, C., G.Y. Akita, W.H. Colledge, R.A. Ratcliff, M.J. Evans, K.M. Hehir, J.A. St. George, S.C. Wadsworth, and S.H. Cheng. 1997. Increased contact time improves adenovirus-mediated CFTR gene transfer to nasal epithelium of CF mice. Hum. Gene Ther. *8*:671-680.

76. Jiang, A., T.-H. Chu, F. Nocken, K. Cichutek, and R. Dornburg. 1998. Cell-type specific gene transfer into human cells with retroviral vectors that display single-chain antibodies. J. Virol. *72*:10148-10156.

77. Johnson, L.G., J.P. Mewshaw, H. Ni, T. Friedmann, R.C. Boucher, and J.C. Olsen. 1998. Effect of host modification and age on airway epithelial gene transfer mediated by a murine leukemia virus-derived vector. J. Virol. *72*:8861-8872.

78. Johnson, L.G., J.C. Olsen, L. Naldini, and R.C. Boucher. 2000. Pseudotyped human lentiviral vector-mediated gene transfer to airway epithelia *in vivo*. Gene Ther. *7*:568-574.

79. Johnson, L.G., J.C. Olsen, B. Sarkadi, K.L. Moore, R. Swanstrom, and R.C. Boucher. 1992. Efficiency of gene transfer for restoration of normal airway epithelial function in cystic fibrosis. Nat. Genet. *2*:21-25.

80. Johnson, L.G., R.J. Pickles, S.E. Boyles, J.C. Morris, H. Ye., Z.Q. Zhou, J.C. Olsen, and R.C. Boucher. 1996. *In vitro* assessment of variables affecting the efficiency and efficacy of adenovirus-mediated gene transfer to cystic fibrosis airway epithelia. Hum. Gene Ther. *7*:51-59.

81. Joos, K., L.A. Turka, and J.M. Wilson. 1998. Blunting of immune responses to adenoviral vectors in mouse liver and lung with CTLA4Ig. Gene Ther. *5*:309-319.

82. Kaplan, J.M., D. Armentano, T.E. Sparer, S.G. Wynn, P.A. Peterson, S.C. Wadsworth, K.K. Couture, S.E. Pennington, J.A. St. George, L.R. Gooding, and A.E. Smith. 1997. Characterization of factors involved in modulating persistence of transgene expression from recombinant adenovirus in the mouse lung. Hum. Gene Ther. *8*:45-56.

83. Kaplan, J.M., J.A. St. George, S.E. Pennington, L.D. Keyes, R.P. Johnson, S.C. Wadsworth, and A.E. Smith. 1996. Humoral and cellular immune responses of nonhuman primates to long-term repeated lung exposure to Ad2/CFTR2. Gene Ther. *3*:117-127.

84. Kaplan, J.M., S.E. Pennington, J.A. St. George, L.A. Woodworth, A. Fasbender, J. Marshall, S.H. Cheng, S.C. Wadsworth, R.J. Gregory, and A.E. Smith. 1998. Potentiation of gene transfer to mouse lung by complexes of adenovirus vector and polycations improves therapeutic potential. Hum. Gene Ther. *9*:1469-1479.

85. Kasahara, N., A.M. Dozy, and Y.W. Kan. 1994. Tissue-specific targeting of retroviral vectors through ligand-receptor interactions. Science *266*:1373-1376.

86. Kass-Eisler, A., L. Leinwand, J. Gall, B. Bloom, and E. Falck-Pedersen. 1996. Circumventing the immune response to adenovirus mediated gene therapy. *3*:154-162.

87. Khan, T.Z., J.S. Wagener, T. Bost, J. Martinez, F.J. Accurso, and D.W. Riches. 1995. Early pulmonary inflammation in infants with cystic fibrosis. Am. J. Resp. Crit. Care Med. *151*:1075-1082.

88. Kirchner, K.K., J.S. Wagener, S.C. Copenhaver, and F.J. Accurso. 1996. Increased DNA levels in bronchoalveolar lavage fluid obtained from infants with cystic fibrosis. Am. J. Resp. Crit. Care Med. *154*:1426-1429.

89. Kitten, O., F.L. Cosset, and N. Ferry. 1997. Highly efficient retrovirus-mediated gene transfer into rat hepatocytes *in vivo*. Hum. Gene Ther. *8*:1491-1494.

90. Klinger, J.R., R.R. Warburton, L.A. Pietras, O. Smithies, R. Swift, and N.S. Hill. 1999. Genetic disruption of atrial natriuretic peptide causes pulmonary hypertension in normoxic and hypoxic mice. Am. J. Physiol. *276*:L868-L874.

91. Knowles, M.R., K. Hohneker, Z.-Q. Zhou, J.C. Olsen, T.L. Noah, P.C. Hu, M.W. Leigh, J.F. Engelhardt, L.J. Edwards, K.R. Jones, et al. 1995. A controlled study of adenoviral vector mediated gene transfer in the nasal epithelium of patients with cystic fibrosis. N. Engl. J. Med. *333*:823-831.

92. Kochanek, S., P.R. Clemens, K. Mitani, H.-H. Chen, S. Chan, and C.T. Caskey. 1996. A new adenoviral vector: replacement of all viral coding sequences with 28 kb of DNA independently expressing both full-length dystrophin and β-galactosidase. Proc. Natl. Acad. Sci. USA *93*:5731-5736.

93. Kollen, W.J.W., F. Schembri, G.J. Gerwig, J.F.G. Vliegenthart, M.C. Glick, and T.F. Scanlin. 1999. Enhanced efficiency of lactosylated poly-L-lysine-mediated gene transfer into cystic fibrosis airway epithelial cells. Am. J. Respir. Cell Mol. Biol. *20*:1081-1086.

94.Kovesdi, I., D.E. Brough, J.T. Bruder, and T.J. Wickham. 1997. Adenoviral vectors for gene transfer. Curr. Opin. Biotechnol. *8*:583-589.

95.Kreda, S.M., R. Pickles, E. Lazarowski, and R.C. Boucher. 1998. P2Y2 receptor agonists to produce conjugates for gene transfer. Ped. Pulmonol. (Suppl.) *17*:258.

96.Ledley, F.D. 1995. Nonviral gene therapy: the promise of genes as pharmaceutical products. Hum. Gene Ther. *6*:1129-1144.

97.Lee, E.R., J. Marshall, C.S. Siegel, C. Jiang, N.S. Yew, M.R. Nichols, J.B. Nietupski, R.J. Ziegler, M.B. Lane, K.X. Wang, et al. 1996. Detailed analysis of structures and formulations of cationic lipids for efficient gene transfer to the lung. Hum. Gene Ther. *7*:1701-1717.

98.Legendre, J.Y. and F.C. Szoka, Jr. 1992. Delivery of plasmid DNA into mammalian cell lines using pH-sensitive liposomes: comparison with cationic liposomes. Pharmaceut. Res. *9*:1235-1242.

99.Leigh, M.W., J.E. Kylander, J.R. Yankaskas, and R.C. Boucher. 1995. Cell proliferation in bronchial epithelium and submucosal glands of cystic fibrosis patients. Am. J. Resp. Cell Mol. Biol. *12*:605-612.

100.Lieber, A., C.Y. He, I. Kirillova, and M.A. Kay. 1996. Recombinant adenoviruses with large deletions generated by cre-mediated excision exhibit different biological properties compared with first-generation vectors *in vitro* and *in vivo*. J. Virol. *70*:8944-8960.

101.Liu, F., H. Qi, L. Huang, and D. Liu. 1997. Factors controlling the efficiency of cationic lipid-mediated transfection *in vivo* via intravenous administration. Gene Ther. *4*:517-523.

102.Look, D.C. and S.L. Brody. 1999. Engineering viral vectors to subvert the airway defense response. Am. J. Respir. Cell Mol. Biol. *20*:1103-1106.

103.Lutz, K.L. and T.J Siahaan. 1997. Molecular structure of the apical junction complex and its contribution to the paracellular barrier. Pharmaceut. Sci. *86*:977984.

104.Mack, C.A., W.R. Song, H. Carpenter, T.J. Wickham, I. Kovesdi, B.G. Harvey, C.J. Magovern, O.W. Isom, T. Rosengart, E. Falck-Pedersen, et al. 1997. Circumvention of anti-adenovirus neutralizing immunity by administration of an adenoviral vector of an alternate serotype. Hum. Gene Ther. *8*:99-109.

105.Mason, S.J., A.M. Paradiso, and R.C. Boucher. 1991. Regulation of transepithelial ion transport and intracellular calcium by extracellular ATP in human normal and cystic fibrosis airway epithelium. Br. J. Pharmacol. *103*:1649-1656.

106.Mastrangeli, A., B.G. Harvey, J. Yao, G. Wolff, I. Kovesdi, R.G. Crystal, and E. Falck-Pedersen. 1996. Seroswitch Ad-mediated *in vivo* gene transfer: circumvention of anti-adenovirus humoral immune defenses against repeat adenovirus vector administration by changing the adenovirus serotype. Hum. Gene Ther. *7*:79-87.

107.Matsui, H., L.G. Johnson, S.H. Randell, and R.C. Boucher. 1997. Loss of binding and entry of liposome-DNA complexes decreases transfection efficiency in differentiated rat tracheal epithelial cells. J. Biol. Chem. *272*:1117-1126.

108.McCray, P.B., Jr., G. Wang, J.N. Kline, J. Zabner, S. Chada, D.J. Jolly, S.M.W. Chang, and B.L. Davidson. 1997. Alveolar macrophages inhibit retrovirus-mediated gene transfer to airway epithelia. Hum. Gene Ther. *8*:1087-1093.

109.McDonald, R.J., M.J. Lukason, O.G. Raabe, D.R. Canfield, E.A. Burr, J.M. Kaplan, S.C. Wadsworth, and J.A. St. George. 1997. Safety of airway gene transfer with Ad2/CFTR2: aerosol administration in the nonhuman primate. Hum. Gene Ther. *8*:411-422.

110.Miller, A.D. and G.J. Rosman. 1989. Improved retroviral vectors for gene transfer and expression. BioTechniques *7*:980-990.

111.Miller, D.G., M.A. Adam, and A.D. Miller. 1990. Gene transfer by retrovirus vectors occurs only in cells that are actively replicating at the time of infection. Mol. Cell. Biol. *10*:4329.

112.Morsy, M.A., M.-C. Gu, S. Motzel, J. Zhao, J. Lin, Q. Su, H. Allen, L. Franlin, R.J. Parks, F.L. Graham, et al. 1998. An adenoviral vector deleted for all viral coding sequences results in enhanced safety and extended expression of a leptin transgene. Proc. Natl. Acad. Sci. USA *95*:7866-7871.

113.Naldini, L. 1998. Lentiviruses as gene transfer agents for delivery to non-dividing cells. Curr. Opin. Biotechnol. *9*:457-463.

114.Naldini, L., U. Blomer, P. Gallay, D. Ory, R. Mulligan, F.H. Gage, I.M. Verma, and D. Trono. 1996. *In vivo* gene delivery and stable transduction of nondividing cells by a lentiviral vector. Science *272*:263-267.

115.Noah, T.L., H.R. Black, P.W. Cheng, R.E. Wood, and M.W. Leigh. 1997. Nasal and bronchoalveolar lavage fluid cytokines in early cystic fibrosis. J. Infect. Dis. *175*:638-647.

116. Noone, P.G., K.W. Hohneker, Z. Zhou, L.G. Johnson, C. Foy, C. Gipson, K. Jones, T.L. Noah, M.W. Leigh, C. Schwartzbach, et al. 2000. Safety and biological efficacy of a lipid-CFTR complex for gene transfer in the nasal epithelium of adult patients with cystic fibrosis. Mol. Ther. *1*:105-114.

117. Olsen, J.C. 1998. Gene transfer vectors derived from equine infectious anemia virus. Gene Ther. *5*:1481-1487.

118. Olsen, J.C., L.G. Johnson, M.J. Stutts, B. Sarkadi, J.R. Yankaskas, R. Swanstrom, and R.C. Boucher. 1992. Correction of the apical membrane chloride permeability defect in polarized cystic fibrosis airway epithelia following retroviral-mediated gene transfer. Hum. Gene Ther. *3*:253-266.

119. O'Riordan, C.R., A. Lachapelle, C. Delgado, V. Parkes, S.C. Wadsworth, A.E. Smith, and G.E. Francis. 1999. PEGylation of adenovirus with retention of infectivity and protection from neutralizing antibody *in vitro* and *in vivo*. Hum. Gene Ther. *10*:1349-1358.

120. Otake, K., D.L. Ennist, K. Harrod, and B.C. Trapnell. 1998. Nonspecific inflammation inhibits adenovirus-mediated pulmonary gene transfer and expression independent of specific acquired immune responses. Hum. Gene Ther. *9*:2207-2222.

121. Paillard, F. 1998. Commentary: cell-specific targeting with retroviral vectors. Hum. Gene Ther. *9*:767-768.

122. Parks, R.J., L. Chen, M. Anton, U. Sankar, M.A. Rudnicki, and F.L. Graham. 1996. A helper-dependent adenovirus vector system: removal of helper virus by cre-mediated excision of the viral packaging signal. Proc. Natl. Acad. Sci. USA *93*:13565-13570.

123. Parsons, D.W., B.R. Grubb, L.G. Johnson, and R.C. Boucher. 1998. Enhanced *in vivo* airway gene transfer via transient modification of host barrier properties with a surface-active agent. Hum. Gene Ther. *9*:2661-2672.

124. Pear, W.D., G. Nolan, M.L. Scott, and D. Baltimore. 1993. Production of high-titer helper-free retroviruses by transient transfection. Proc. Natl. Acad. Sci. USA *90*:8392-8396.

125. Perlmutter, A.H. 1996. Alpha 1-antirypsin deficiency: biochemistry and clinical manifestations. Ann. Med. *28*:385-394.

126. Pickles, R.J., P.M. Barker, H. Ye, and R.C. Boucher. 1996. Efficient Ad-mediated gene transfer to basal but not columnar cells of cartilaginous airway epithelia. Hum. Gene Ther. *7*:921-931.

127. Pickles, R.J., L.G. Johnson, J.C. Olsen, R. Gerard, D. Segal, and R.C. Boucher. 1999. Correction of the CF bioelectrical defect in human CF well differentiated airway epithelail cells by re-targeting adenoviral vectors to luminal P2Y2 purinoceptors [abstract]. Ped. Pulmonol. (Suppl.) *19*:222.

128. Pickles, R.J., S. Kreda, J. Olsen, L. Johnson, R. Gerard, D. Segal, and R.C. Boucher. 1998. High efficiency gene transfer to polarised epithelial cells by re-targeting adenoviral vectors to P2Y$_2$ purinoceptors with bi-specific antibodies. Ped. Pulmonol. (Suppl.) *17*:261.

129. Pickles, R.J., D. McCarty, H. Matsui, P.J. Hart, S.H. Randell, and R.C. Boucher. 1998. Limited entry of adenovirus vectors into well-differentiated airway epithelium is responsible for inefficient gene transfer. J. Virol. *72*:6014-6023.

130. Piedimonte, G., R.J. Pickles, J.R. Lehmann, D. McCarty, D.L. Costa, and R.C. Boucher. 1997. Replication-deficient adenoviral vector for gene transfer potentiates airway neurogenic inflammation. Am. J. Respir. Cell Mol. Biol. *16*:250-258.

131. Pilewski, J.M., J.F. Engelhardt, J.E. Bavaria, L.R. Kaiser, J.M. Wilson, and S.M. Albelda. 1995. Ad-mediated gene transfer to human bronchial submucosal glands using xenografts. Am. J. Physiol. *268*:L657-L665.

132. Porteous, D.J., J.R. Dorin, G. McLachlan, H. Davidson-Smith, H. Davidson, B.J. Stevenson, A.D. Carothers, W.A. Wallace, S. Moralee, C. Hoenes, et al. 1997. Evidence for safety and efficacy of DOTAP cationic liposome mediated CFTR gene transfer to the nasal epithelium of patients with cystic fibrosis. Gene Ther. *4*:210-218.

133. Qing, K., C. Mah, J. Hansen, S. Zhou, V. Dwarki, and A. Srivastava. 1999. Human fibroblast growth factor receptor 1 is a co-receptor for infection by adeno-associated virus 2. Nat. Med. *5*:71-77.

134. Rabinowitz, J.E. and J. Samulski. 1998. Adeno-associated virus expression systems for gene transfer. Curr. Opin. Biotechnol. *9*:470-475.

135. Rich, D.P., M.P. Anderson, R.J. Gregory, S.H. Cheng, S. Paul, D.M. Jefferson, J.D. McCann, K.W. Klinger, A.E. Smith, and M.J. Welsh. 1990. Expression of the cystic fibrosis transmembrane conductance regulator corrects defective chloride channel regulation in cystic fibrosis airway epithe-

lial cells. Nature *347*:358-363.

136. Rich, D.P., L.A. Couture, L.M. Cardoza, V.M. Guiggio, D. Armentano, P.C. Espino, K. Hehir, M.J. Welsh, A.E. Smith, and R.J. Gregory. 1993. Development and analysis of recombinant adenoviruses for gene therapy of cystic fibrosis. Hum. Gene Ther. *4*:461-476.

137. Rich, S. 1998. Primary pulmonary hypertension, p. 1466-1468. *In* A.S. Fauci, E. Braunwald, K.J. Isselbacher, J.D. Wilson, J.B. Martin, D.L. Kasper, S.L. Hauser, and D.L. Longo (Eds.), Harrison's Principles of Internal Medicine. McGraw-Hill, New York.

138. Rosenfeld, M.A., K. Yoshimura, B.C. Trapnell, K. Yoneyama, E.R. Rosenthal, W. Dalemans, M. Fukayama, J. Bargon, L.E. Stier, L. Stratford-Perricaudet, et al. 1992. *In vivo* transfer of the human cystic fibrosis transmembrane conductance regulator gene to the airway epithelium. Cell *68*:143-155.

139. Roth, J.A., D. Nguyen, D.D. Lawrence, B.L. Kemp, C.H. Carrasco, D.Z. Ferson, W.K. Hong, R. Komaki, J.J. Lee, J.C. Nesbitt, et al. 1996. Retrovirus-mediated wild-type p53 gene transfer to tumors of patients with lung cancer. Nat. Med. *2*:985-991.

140. Roth, J.A., S.G. Swisher, J.A. Merritt, D.D. Lawrence, B.L. Kemp, C.H. Carrasco, A.K. El-Naggar, F.V. Fossella, B.S. Glisson, W.K. Hong, et al. 1998. Gene therapy for non-small cell lung cancer: a preliminary report of a phase I trial of adenoviral p53 gene replacement. Sem. Oncol. (3 Suppl. 8) *25*:33-37.

141. Russell, D.W., I.E. Alexander, and A.D. Miller. 1995. DNA synthesis and topoisomerase inhibitors increase transduction by adeno-associated virus vectors. Proc. Natl. Acad. Sci. USA *92*:5719-5723.

142. Scaria, A., J.A. St. George, R.J. Gregory, R.J. Noelle, S.C. Wadsworth, A.E. Smith, and J.M. Kaplan. 1997. Antibody to CD40 ligand inhibits both humoral and cellular immune responses to adenoviral vectors and facilitates repeated administration to mouse airway. Gene Ther. *4*:611-617.

143. Scherman, D., M. Bessodes, B. Cameron, J. Herscovici, H. Hofland, B. Pitard, F. Soubrier, P. Wils, and J. Crouzet. 1998. Application of lipids and plasmid design for gene delivery to mammalian cells. Curr. Opin. Biotechnol. *9*:480-485.

144. Scheule, R.K., J.A. St. George, R.G. Bagley, J. Marshall, J.M. Kaplan, G.Y. Akita, K.X. Wang, E.R. Lee, D.J. Harris, C. Jiang, et al. 1997. Basis of pulmonary toxicity associated with cationic lipid-mediated gene transfer to the mammalian lung. Hum. Gene Ther. *8*:689-707.

145. Schierle, B.S., D. Moritz, M. Jeschke, and B. Groner. 1996. Expression of chimeric envelope proteins in helper cell lines and integration into moloney murine leukemia virus particles. Gene Ther. *3*:334342.

146. Schreier, H., P. Moran, and I.W. Caras. 1994. Targeting of liposomes to cells expressing CD4 using glycosylphosphatidylinositol-anchored gp120. J. Biol. Chem. *12*:9090-9098.

147. Schwarzenberger, P., S.E. Spence, J.M. Gooya, D. Michiel, D.T. Curiel, F.W. Ruscetti, and J.R. Keller. 1996. Targeted gene transfer to human hematopoietic progenitor cell lines through the c-kit receptor. Blood *87*:472-478.

148. Simon, R.H., J.F. Engelhardt, Y. Yang, M. Zepeda, S. Weber-Pendleton, M. Grossman, and J.M. Wilson. 1993. Adenovirus-mediated transfer of the CFTR gene to lung of non-human primates: toxicity study. Hum. Gene Ther *4*:771-780.

149. Soneoka, Y., P.M. Cannon, E.F. Ramsdale, J.C. Griffiths, G. Romano, S.M. Kingsman, and A.J. Kingsman. 1995. A transient three-plasmid system for the production of high titer retroviral vectors. Nucleic Acids Res. *23*:628-633.

150. St. George, J.A., S.E. Pennington, J.M. Kaplan, P.A. Peterson, L.J. Kleine, and A.E. Smith. 1996. Biological response of nonhuman primates to long-term repeated lung exposure to Ad2/CFTR-2. Gene Ther. *3*:103-116.

151. Stern, M., N.J. Caplen, J.E. Browning, U. Griesenbach, F. Sorgi, L. Huang, and D.C. Gruenert. 1998. The effect of mucolytic agents on gene transfer across a CF sputum barrier *in vitro*. Gene Ther. *5*:91-98.

152. Summerford, C., J.S. Bartlett, and R.J. Samulski. 1999. $\alpha_V\beta_5$ integrin: a co-receptor for adeno-associated virus type 2 infections. Nat. Med. *5*:78-81.

153. Summerford, C. and R.J. Samulski. 1998. Membrane-associated heparan sulfate proteoglycan is a receptor for adeno-associated virus type 2 virions. J. Virol. *72*:1438-1445.

154. Summerford, C. and R.J. Samulski. 1999. Viral receptors and vector purification: new approaches for generating clinical-grade reagents. Nat. Med. *5*:587-588.

155. Swisher S.G., J.A. Roth, J. Nemunaitis, D.D. Lawrence, B.L. Kemp, C.H. Carrasco, D.G. Con-

nors, A.K. El-Naggar, F. Fossella, B.S. Glisson, et al. 1999. Adenovirus-mediated p53 gene transfer in advanced non-small-cell lung cancer. J. Natl. Cancer Inst. *91*:763-771.

156. Tepper, R.S., G.L. Montgomery, V. Ackerman, and H. Eigen. 1993. Longitudinal evaluation of pulmonary function in infants and very young children with cystic fibrosis. Ped. Pulmonol. *16*:96-100.

157. Teramoto, S., L.G. Johnson, W. Huang, M.W. Leigh, and R.C. Boucher. 1995. Effect of adenoviral vector infection on cell proliferation in a cultured primary human airway epithelial cells. Hum. Gene Ther. *6*:1045-1053.

158. Tuder, R.M., C.D. Cool, M.W. Geraci, J. Wang, S.H. Abman, L. Wright, D. Badesch, and N.F. Voelkel. 1999. Prostacyclin synthase expression is decreased in lungs from patients with severe pulmonary hypertension. Am. J. Respir. Crit. Care Med. *159*:1925-1932.

159. Van Ginkel, F.W., C.-G. Liu, J.W. Simecka, J.Y. Dong, T. Greenway, R.A. Frizzell, H. Kiyono, J.R. McGhee, and D.W. Pascual. 1995. Intratracheal gene delivery with adenoviral vector induces elevated systemic IgG and mucosal IgA antibodies to adenovirus and β-galactosidase. Hum. Gene Ther. *6*:895-903.

160. Van Heeckeren, A., T. Ferkol, and M. Tosi. 1998. Effects of bronchopulmonary inflammation induced by pseudomonas aeruginosa on adenovirus-mediated gene transfer to airway epithelial cells in mice. Gene Ther. *5*:345-351.

161. Voelkel, N.F., R.M. Tuder, K. Wade, M. Hoeper, R.A. Lepley, J.L. Goulet, B.H. Koller, and F. Fitzpatrick. 1996. Inhibition of 5-lipoxygenase-activating protein (FLAP) reduces pulmonary vascular reactivity and pulmonary hypertension in hypoxic rats. J. Clin. Invest. *97*:2491-2498.

162. Wagner, J.A., T. Reynolds, M.L. Moran, R.B. Moss, J.J. Wine, T.R. Flotte, and P. Gardner. 1998. Efficient and persistent gene transfer of AAV-CFTR in maxillary sinus. Lancet *351*:1702-1703.

163. Walters, R.W., T. Grunst, J.M. Bergelson, R.W. Finberg, M.J. Welsh, and J. Zabner. 1999. Basolateral localization of fiber receptors limits adenovirus infection from the apical surface of airway epithelia. J. Biol. Chem. *274*:10219-10226.

164. Walters, R.W., D.-S. Duan, J.F. Engelhardt, and M.J. Welsh. 2000. Incorporation of adeno-associated virus in a calcium phosphate coprecipitate improves gene transfer to airway epithelia *in vitro* and *in vivo*. J. Virol. *74*:535-540.

165. Wang, G., B.L. Davidson, P. Melchert, V.A. Slepushkin, H.H.G. van Es, M. Bodner, D.J. Jolly, and P.B. McCray, Jr. 1998. Influence of cell polarity on retrovirus-mediated gene transfer to differentiated human airway epithelia. J. Virol. *72*:9818-9826.

166. Wang, G., V.A. Slepushkin, M. Bodner, J. Zabner, H.H.G. van Es, P. Thomas, D.J. Jolly, B.L. Davidson, and P.B. McCray, Jr. 1999. Keratinocyte growth factor induced epithelial proliferation facilitates retroviral-mediated gene transfer to distal lung epithelial *in vivo*. J. Gene Med. *1*:22-30.

167. Wang, G., V.A. Slepushkin, J. Zabner, S. Keshavjee, J.C. Johnston, S.L. Sauter, D.J. Jolly, T.W. Dubensky, Jr., B.L. Davidson, and P.B. McCray, Jr. 1999. Feline immunodeficiency virus vectors persistently trtransduce nondividing airway epithelia and correct the cystic fibrosis defect. J. Clin. Invest. *104*:R55-R62.

168. Wang, Q., X.-C. Jia, and M.H. Finer. 1995. A packaging cell line for propagation of recombinant adenovirus vectors containing two lethal gene-region deletions. Gene Ther. *2*:775-783.

169. Watkins, S.J., V.V. Mesyanzhinov, L.P. Kurochkina, and R.E. Hawkins. 1997. The adenobody approach to viral targeting: specific and enhanced adenoviral gene delivery. Gene Ther. *4*:1004-1012.

170. Weiss, D.J., T.P. Strandjord, D. Liggitt, and J.G. Clark. 1999. Perflubron enhances adenovirus-mediated gene expression in lungs of transgenic mice with chronic alveolar filling. Hum. Gene Ther. *10*:2287-2293.

171. Welsh, M.J., J. Zabner, S.M. Graham, A.E. Smith, R. Moscicki, and S. Wadsworth. 1995. Adenovirus-mediated gene transfer for cystic fibrosis: Part A Safety of dose and repeat administration in the nasal epithelium. Part B. Clinical efficacy in the maxillary sinus. Hum. Gene Ther. *6*:205-218.

172. Wickham, T.J., G.M. Lee, J.A. Titus, G. Sconocchia, T. Bakacs, I. Kovesdi, and D.M. Segal. 1997. Targeted adenovirus-mediated gene delivery to T cells via CD3. J. Virol. *71*:7663-7669.

173. Wickham, T.J., P. Mathias, D.A Cheresh, and G.R. Nemerow. 1993. Integrins $\alpha_v\beta_3$ and $\alpha_v\beta_5$ promote adenovirus internalization but not virus attachment. Cell *73*:309-319.

174. Wickham, T.J., E. Tzeng, L.L. Shears II, P.W. Roelvink, Y. Li, G.M. Lee, D.E. Brough, A. Lizonova, and I. Kovesdi. 1997. Increased *in vitro* and *in vivo* gene transfer by adenovirus vectors containing chimeric fiber proteins. J. Virol. *71*:8221-8229.

175. Wiedemann, H.P. and J.K. Stoller. 1996. Lung disease due to alpha-1-antitrypsin. Curr. Opin. in Pulm. Med. *2*:155-160.

176. Wilson, C.B., L.J. Embree, D. Schowalter, R. Albert, A. Aruffo, D. Hollenbaugh, P. Linsley, and M.A. Kay. 1998. Transient inhibition of CD28 and CD40 ligand ineractions prolongs adenovirus-mediated transgene expression in the lung and facilitates expression after secondary vector administration. J. Virol. *72*:7542-7550.

177. Worgall, S., P.L. Leopold, G. Wolfe, B. Ferris, N. Van Roijen, and R.G. Crystal. 1997. Role of alveolar macrophages in rapid elimination of adenovirus vectors administered to the epithelial surface of the respiratory tract. Hum. Gene Ther. *8*:1675-1684.

178. Xiao, X., J. Li, and R.J. Samulski. 1998. Production of high titer recombinant adeno-associated virus vectors in the absence of helper virus. J. Virol. *72*:2224-2232.

179. Yajima, T., T. Kanda, K. Yoshiike, and Y. Kitamura. 1998. Retroviral vector targeting human cells via c-kit-stem cell factor interaction. Hum. Gene Ther. *9*:779-788.

180. Yang, Y., Q. Li, H.C.J. Ertl, and J.M. Wilson. 1995. Cellular and humoral immune responses to viral antigens create barriers to lung-directed gene therapy with recombinant adenoviruses. J. Virol. *69*:2004-2015.

181. Yang, Y., F.A. Nunes, K. Berencsi, E. Gonczol, J.F. Engelhardt, and J.M. Wilson. 1994. Inactivation of E2a in recombinant adenoviruses limits cellular immunity and improves the prospect for gene therapy of cystic fibrosis. Nat. Genet. *7*:362-369.

182. Yang, Y., G. Trinchieri, and J.M. Wilson. 1995. Recombinant IL-12 prevents formation of blocking IgA antibodies to recombinant adenovirus and allows repeated gene therapy to mouse lung. Nat. Med. *1*:890-893.

183. Yei, S., N. Mittereder, S. Wert, J.A. Whitsett, R.W. Wilmott, and B.C. Trapnell. 1994. *In vivo* evaluation of the safety of adenovirus-mediated transfer of the human cystic fibrosis transmembrane conductance regulator cDNA to the lung. Hum. Gene Ther. *5*:731-744.

184. Yei, S., N. Mittereder, K. Tang, C. O'Sullivan, and B.C. Trapnell. 1994. Adenovirus-mediated gene transfer for cystic fibrosis: quantitative evaluation of repeated *in vivo* vector administration to the lung. Gene Ther. *1*:192-200.

185. Zabner, J., S.H. Cheng, D. Meeker, J. Launspach, R. Balfour, M.A. Perricone, J.E. Morris, J. Marshall, A. Fasbender, A.E. Smith, and M.J. Welsh. 1997. Comparison of DNA-lipid complexes and DNA alone for gene transfer to cystic fibrosis airway epithelia *in vivo*. J. Clin. Invest. *100*:1529-1537.

186. Zabner, J., M. Chillon, T. Grunst, T.O. Moninger, B.L. Davidson, R. Gregory, and D. Armentano. 1999. A chimeric type 2 adenovirus vector with a type 17 fiber enhances gene transfer to human airway epithelia. J. Virol. *73*:8689-8695.

187. Zabner, J., L.A. Couture, R.J. Gregory, S.M. Graham, A.E. Smith, and M.J. Welsh. 1993. Ad-mediated gene transfer transiently corrects the chloride transport defect in nasal epithelia of patients with cystic fibrosis. Cell *75*:207-216.

188. Zabner, J., L.A. Couture, A.E. Smith, and M.J. Welsh. 1994. Correction of cAMP-stimulated fluid secretion in cystic fibrosis airway epithelia: efficiency of Ad-mediated gene transfer *in vitro*. Hum. Gene Ther. *5*:585-593.

189. Zabner, J., A.J. Fasbender, T. Moninger, K.A. Poellinger, and M.J. Welsh. 1995. Cellular and molecular barriers to gene transfer by a cationic lipid. J. Biol. Chem. *270*:18997-19007.

190. Zabner, J., P. Freimuth, A. Puga, A. Fabrega, and M.J. Welsh. 1997. Lack of high affinity fiber receptor activity explains the resistance of ciliated airway epithelia to adenovirus infection. J. Clin. Invest. *100*:1144-1149.

191. Zabner, J., B.W. Ramsey, D.P. Meeker, M.L. Aitken, R.P. Balfour, R.L. Gibson, J. Launspach, R.A. Moscicki, S.M. Richards, T.A. Standaert, et al. 1996. Repeat administration of an adenovirus vector encoding cystic fibrosis transmembrane conductance regulator to the nasal epithelium of patients with cystic fibrosis. J. Clin. Invest. *97*:1504-1511.

192. Zabner, J., S.C. Wadsworth, A.E. Smith, and M.J. Welsh. 1996. Adenovirus-mediated generation of cAMP-stimulated Cl-transport in cystic fibrosis airway epithelia *in vitro*: effect of promoter and administration method. Gene Ther. *3*:458-465.

193. Zabner, J., B.G. Zeiher, E. Friedman, and M.J. Welsh. 1996. Adenovirus-mediated gene transfer to ciliated airway epithelia requires prolonged incubation time. J. Virol. *70*:6994.

194. Zeitlin, P.L. 1997. Adeno-associated virus-based delivery systems, p. 53-81. *In* K. Brigham (Ed.),

Lung Biology in Health and Disease: Gene Therapy for Diseases of the Lung. Marcel Dekker, New York.

195. Zhou, X. and L. Huang. 1994. DNA transfection mediated by cationic liposomes containing lipopolylysine: characterization and mechanism of action. Biochim. Biophys. Acta *1189*:195-203.

196. Ziady, A.G., R. Keefer, T. Ferkol, and P.B. Davis. 1999. Serpin enzyme complex receptor targeted DNA complexes deliver genes to airway epithelia. Ped. Pulmonol. (Suppl.) *19*:233.

197. Zolotukhin, S., B.J. Byrne, E. Mason, I. Zolotukhin, M. Potter, K. Chesnut, C. Summerford, R.J. Samulski, and N. Muzyczka. 1999. Recombinant adeno-associated virus purification using novel methods improves infectious titer and yield. Gene Ther. *6*:973-985.

198. Zsengeller, Z.K., C. Halbert, A.D. Miller, S.E. Wert, J.A. Whitsett, and C.J. Bachurski. 1999. Keratinocyte growth factor stimulates transduction of the respiratory epithelium by retroviral vectors. Hum. Gene Ther. *10*:341-353.

199. Zufferey, R., T. Dull, R.J. Mandel, A. Bukovsky, D. Quiroz, L. Naldini, and D. Trono. 1998. Self-inactivating lentivirus vector for safe and efficient *in vivo* gene delivery. J. Virol. *72*:9873-9880.

200. Zufferey, R., D. Nagy, R.J. Mandel, L. Naldini, and D. Trono. 1997. Multiply attenuated lentiviral vector achieves efficient gene delivery *in vivo*. Nat. Biotechnol. *15*:871-875.

10 Genetic Manipulation and the Precautionary Principle

Stuart A. Newman[1] and Carolyn Raffensperger[2]
[1]*Department of Cell Biology and Anatomy, New York Medical College, Valhalla, NY;* [2]*Science and Environmental Health Network, Windsor, ND, USA*

INTRODUCTION

The various gene-based organism modification methods described in this volume, whatever benefits may ultimately result from them, carry an array of risks that are in many cases poorly understood or entirely uncharacterized. The customary response of scientists when confronted with such uncertainties is to perform the studies and analyses required to dispel them—to do the experiment. However, the techniques discussed here can potentially lead to new microbial pathogens, weeds, and insect pests, environmental disruption, as well as to adverse health effects in human individuals undergoing somatic gene manipulation, and to changes in the human germline. Clearly it is inappropriate to treat as experimental systems the biosphere and its component ecosystems, or the human species and its gene pool.

Another deep-seated attitude among working scientists is skepticism about the existence of an effect in the absence of strong evidence for it. The standard of proof for the rejection of a hypothesis of no effect (the null hypothesis) is high. In statistical terms, it is generally characterized by a criterion in which the reported effect must have a less than 1 in 20 chance of having arisen by pure chance. This simply stated statistical standard conceals a complex framework of assumptions specific to each field, concerning what constitutes an appropriate set of controls, which factors can reasonably be held constant without invalidating the experiment, what constitutes legitimate reasons for discarding data, what "goes without saying," etc. Despite these complexities, the high standard of proof for the finding of an effect has served

Gene Transfer Methods
Edited by Pamela A. Norton and Laura F. Steel
©2000 Eaton Publishing, Natick, MA

231

many branches of science well, helping to guard against the proliferation of unfounded assertions. In particular, it has created an asymmetry in which it is considered better science to erroneously claim that there is no effect than to erroneously claim that there is an effect (5).

Errors are errors though, however tolerable some categories may be in the culture of scientific investigation. But it should also be clear that what works for the testing of narrowly posed scientific hypotheses is not necessarily a good guide for decision making in other arenas. A hiker confronted with data that two of her ten companions have become severely ill after eating an unfamiliar type of mushroom will probably not apply scientifically standard criteria of statistical significance in rejecting the null hypothesis. Similarly, the potentially deleterious effects of chemicals and drugs on the developing embryo (even including moderately elevated levels of essential nutrients, such as vitamin A) makes it prudent to avoid recreational substances and medications during pregnancy, although the latter is not always feasible. Anecdotal case reports in Germany and Australia in the 1960s alerted physicians to the harmful effects on embryonic development of the sleeping pill thalidomide years before any statistically sound epidemiology was available. In the face of protestations by the manufacturer that nothing had been proved, a heroically cautious FDA official kept the drug from being approved for wide use in the U.S. Accordingly, what serves science well (reluctance to reject the null hypothesis), may not serve public policy when there are high stakes if the null hypothesis has wrongly been accepted and there is indeed an effect causally linked to the chemical or process under investigation, which the experiment has not uncovered.

The examples given above are characterized by there being more at stake than the outcome of a scientific experiment. Of course, this is not only an issue for advocates of precaution in implementing new technologies. Those who wish to accelerate such applications often point to the costs of not curing cancer, AIDS, world hunger, and so forth, and of course such costs are real. In the case of proponents of rapid applications however, there has also been a tendency to stigmatize their critics as antiscience and enemies of free inquiry. This stratagem makes three tacit, interconnected assumptions: *(i)* that applications of technology are the same thing as scientific research, *(ii)* that the environment and the human population are appropriate experimental systems, and *(iii)* that the problems in question are solvable by technological applications of the sorts proposed. The first of these assumptions is false, the second ethically unsound, and while the third may or may not be true, given the often uncritical technological imperative of our culture, it is frequently coercive or misleading.

While we thus reject the identification of applications of research findings (which in any case, are increasingly commercially driven) with freedom of inquiry, we do not deny that some applications can be beneficial, and will inevitably include an experimental component. It is our purpose in this chapter to suggest a framework for considering situations in which the risks are plausible but unquantifiable and where the costs to health and the environment of their occurring would be high.

Our approach is informed by the "precautionary principle," originally codified in the Hippocratic Oath traditionally administered to physicians at the end of their preclinical training: "First do no harm." This theme was elaborated upon by legal scholars early in the twentieth century in Germany in the form of the doctrine of *Vorsorgeprinzip*—"foresight planning"—which encompassed notions of risk prevention, ethical responsibilities towards maintaining the integrity of natural systems, and the fallibility of human understanding (6).

A modern statement of the principle was adopted by the vast majority of U.N. member countries as part of the 1992 Rio Declaration on Environment and Development:

> Where there are threats of serious or irreversible damage, lack of scientific certainty shall not be used as a reason for postponing cost-effective measures to prevent environmental degradation. (U.N. Agenda 21, 1992, 10).

Although the various elements of this statement are subject to differing interpretations (What is serious damage? How is lack of scientific certainty assessed or quantified? What criteria enter into cost-effectiveness?) and other versions of the principle have been proposed (12), the underlying theme in all these statements is a shifting of the burden of proof from those who would forestall or prohibit a technological activity to those who would implement it. In particular, for technologies for which there are plausible threats of irreversible harm, and which have not yet been implemented, or which can feasibly be halted (e.g., the gene modification methods discussed in this volume, in contrast to highly entrenched nongenetic technologies such as extraction and purification of oil or heavy metals), a common interpretation of cost-effective measures to prevent environmental degradation among advocates of the precautionary principle might be simply not to do it. While any good decision-rule should permit saying "yes" or "no" depending on the circumstances, it is unfortunate that current decision rules applied in many western countries based on risk assessment and cost-benefit analysis have only been used to justify new technologies.

A reflexive identification of technological activity with science may cause the precautionary principle to be perceived as in opposition to the scientific method as it has been developed since the Renaissance. But a more nuanced view of rigorous scientific activity in the realm of health and the environment includes consideration of multidisciplinary approaches with incommensurate standards of proof, appreciation of complexity, and inevitable indeterminacy (5). Under such circumstances, precaution indeed becomes a component of the scientific method.

In the following sections, we describe three distinct examples of genetic modification of organisms in which plausible hazards exist: *(i)* the genetic modification of microorganisms for experimental or commercial purposes; *(ii)* the production of genetically modified crops and their introduction into the environment and food chain; and *(iii)* the genetic modification of humans at the embryonic or fetal stages with the expectation (whether or not it is the goal of the modification) that the germline will be affected. These examples are not meant to be exhaustive with regard to the possibilities of genetic manipulation of organisms, but rather paradigmatic, as we will argue that application of the precautionary principle suggests different approaches to each technology, ranging from proceeding with great caution, to not attempting implementation at all. Since our particular conclusions will inevitably be subject to debate, and may be disputed even by those who share our philosophical framework, we will attempt to move beyond specific cases and conclude with a discussion of the differences between an approach informed by the precautionary principle and that adhering to standard risk assessment and cost-benefit analysis.

THREE HAZARDOUS GENE MODIFICATION TECHNOLOGIES

Microorganisms Containing Animal Genes

Little has been written about the hazards of the dissemination of microorganisms containing animal genes since the recombinant DNA debates of the 1970s. At that time, there were speculations concerning the creation of new pathogens by the insertion of animal transgenes into otherwise innocuous bacteria or viruses (72). These scenarios were based on examples in which pathogenicity of naturally occurring microbes was found to result from products of genes, usually dispensable to the microorganism in question, which mimicked, or in the case of enzymes, targeted and modified, certain human proteins. The possibility of producing new pathogens by these routes had essentially no impact on the deliberations that led to the regulation by the National Institutes of Health (NIH) of recombinant DNA research, and the subsequent deregulation of much of it. This was because an early consensus was reached in the relevant working groups [despite evidence to the contrary (2,47,69,83)], that genes cloned into laboratory strains of *Escherichia coli* could not plausibly be transferred into wild strains of bacteria (55,85).

With the advent of monoclonal antibody and gene isolation and sequencing technologies, the existence of molecular mimicry as a mechanism of pathogenesis, particular in the etiology of autoimmune disease, has been amply confirmed. It is now known that both antibodies and T cells are capable of responding to molecular mimicry between pathogens and self antigens in fashions than can eventually lead to disruption of immune tolerance and development of autoimmune disease (13,18,61).

The autoimmune disease for which molecular mimicry is best established as a mechanism of pathogenesis is rheumatic fever, a sequela of group A streptococcal pharyngitis (15). Mimicry between cardiac myosins and tropomyosins, and the streptococcal carbohydrate and M protein, is the most likely etiology of the disease (5,14,16,21,33,40,51). Other conditions believed to be induced by molecular mimicry include insulin-dependent diabetes mellitus (3,78), myasthenia gravis (53,70), and multiple sclerosis (27,87). Mimicry between the HLA-B27 class I molecule and the microbial nitrogenase is believed to play a role in ankylosing spondylitis, arthritis, and Reiter syndrome, which are sequelae of bacterial infections (71). Heat-shock proteins of the intracellular pathogen *Mycobacterium tuberculosis*, which resemble those of the host cell, have been implicated in autoimmune diabetes and arthritis (11). In a transgenic animal model in which mice were caused to express a viral protein in oligodendrocytes, infection with the virus led to chronic central nervous system autoimmune disease (22).

Microorganism-encoded enzymes that modify, and thereby change the function of physiologically essential host proteins, represent another scenario by which microbial transgenesis could lead to novel pathogens (50,84). The diphtheria *tox* gene is encoded by a bacteriophage that renders innocuous *Corynebacterium diphtheriae* pathogenic. Fragment A of the diphtheria toxin inhibits host cell translation by ADP-ribosylating eukaryotic elongation factor 2 (EF-2), as does *Pseudomonas* toxin A. Animal cells contain a mono (ADP-ribosyl) transferase with the same specificity for EF-2 as these bacterial enzymes (36,45), and this cellular enzyme plays a role in the regulation of protein synthesis (35). Cholera toxin (CTX), and a closely related heat-labile enterotoxin produced by *E. coli*, which produces a milder form of diarrhea, are also mono (ADP-ribosyl) transferases, which in these cases act on animal cell membrane proteins, leading to an increase in intracellular cyclic AMP (73). Like diphtheria and *Pseudomonas* toxins, the CTX type of enzyme activity has counterparts in the animal cell (37,60). And, although it was not recognized until very recently, the genes encoding the cholera toxin can be transferred horizontally among otherwise nonpathogenic strains of *Vibrio cholerae* by a filamentous bacteriophage (46,80).

It is plausible that pathogenic molecular mimics and protein modification enzymes were acquired by bacteria, bacteriophages, or viruses by horizontal transfer of DNA from eukaryotes during the course of evolution (7). But several features of modern gene transfer technologies ensure that random horizontal transfer of DNA from animal cells to microorganisms, which only rarely would have led to significant transgene protein expression over the course of evolution, can now do so with much greater frequency (29): *(i)* many laboratory, medical, and commercial applications of genetic technologies are specifically directed to propagating animal DNA in bacteria and viruses, and achieving high expression of the corresponding proteins by use of sophisticated plasmid and viral vectors; *(ii)* the laboratory *E. coli* strains used to propagate recombinant DNA molecules in research and industrial settings readily exchange DNA with nonlaboratory strains of bacteria (2,47,69,83); *(iii)* deregulation by the NIH of research using *E. coli* K12-based host-vector systems has led to laxity concerning what gets released into the environment from these activities.

In addition to molecular mimicry and host protein targeted enzymes, synergistic toxic effects induced by bacteria and their recombinant products can occur and may, in certain situations, enhance the intrinsic toxic capacity of the genetically modified microorganism. For example, interferon (IFN)-gamma-producing genetically modified *E. coli* K12, or mixtures of *E. coli* cells and IFN-gamma, were found to be cytolytic for mouse embryo fibroblastoid cells (MEF), whereas no cell killing occurred in MEF cultures treated with control *E. coli* cells or in those treated with bacteria-free recombinant IFN-gamma (19).

These considerations virtually ensure that genetic technologies in current use are causing otherwise innocuous bacteria and viruses outside the laboratory to express enzymes and other proteins of animal origin, and that the capacity of the microbes to cause disease may thereby be enhanced. It must be acknowledged that no new pathogen has emerged since the start of gene splicing experiments, in the 1970s, that looks to have been caused by transfer of animal genes to microbes. However, a quarter century elapsed between the mass administration in Africa of oral polio vaccines produced in primary cell cultures derived from nonhuman primates and the identification of AIDS as a clinical entity, and evidence is now accumulating that there may have been a connection between these two events (32). Nonetheless, given the scientific, and more to the point, the economic investment in biotechnologies involving transfer of the capacity to make animal proteins to bacteria and viruses, it is impractical to expect cessation of these activities.

The precautionary principle dictates several measures under these circumstances. First, the principle would require an inventory of potential harms (including worst case scenarios) resulting from each modified organism. Second, routine releases of even those genetically modified organisms that have been exempted from the NIH guidelines for the conduct of recombinant DNA experiments (58) would be prohibited, and a vigorous monitoring program of any inadvertent releases of such organisms into the environment, either through the air and laboratory vents, but particularly into the effluents from laboratories and companies involved in gene splicing technologies would be instituted. The monitoring program would be incorporated into the overall research design and undertaken by independent, third-party scientists. Third, emergency and disaster plans would be shared with government officials and local communities with detailed planning for appropriate and rapid response to outbreaks, based on the reasonable expectation that new pathogens will eventually arise from these activities. Fourth, companies would post performance bonds that would be forfeited if an unplanned release and/or damage occurred either during the research phase or after commercial release.

Genetically Modified Crops

There are several kinds of risk associated with the genetic modification of crops that make application of the precautionary principle relevant. The main issues are the environmental effects of herbicide- and pesticide-resistant crops, and the effects on human health of ingestion of food containing transgenes. The environmental issues have been reviewed by Krimsky and Wrubel (41), Rissler and Mellon (67), and Altieri (1), who concluded that the most serious ecological risks posed by the commercial-scale use of transgenic crops are:

- The introduction of genes for herbicide resistance can potentially transform a crop plant into a weed.
- The sexual transfer of genes from herbicide-resistant crops to wild or semidomesticated relatives can potentially create new weeds.
- Vector-mediated horizontal gene transfer and recombination may create new pathogenic bacteria.
- Vector recombination may generate new virulent strains of virus, especially in transgenic plants engineered for viral resistance with viral genes.
- Insect pests will quickly develop resistance to crops engineered to express *Bacillus thuringiensis* (Bt) or other transgenic toxins.
- Massive use of transgenic toxins in crops can unleash potential negative interactions affecting ecological processes and nontarget organisms.
- The spread of transgenic crops threatens crop genetic diversity, making the food supply susceptible to a few pests or diseases.

Readers interested in detailed analysis of these issues can consult the works cited and the references therein. A few relevant points will be discussed briefly here. Proponents of herbicide-resistant crops claim that this technology enables farmers to simplify weed management by reducing herbicide use to post-emergence situations using a single, broad-spectrum herbicide that breaks down relatively rapidly in the soil. Glyphosate (Roundup; Monsanto/Pharmacia, Peapack, NJ, USA) is a widely used herbicide with these characteristics, and the majority of genetically modified crops that have been brought to market to date are glyphosate-resistant.

With the repeated use of a given herbicide on a crop, the chances of resistance developing in associated weed populations greatly increases (31). Cocklebur, an aggressive weed of soybean and corn in the southeastern U.S., has become resistant to imidazolinone herbicides (26), and resistance by weeds to glyphosate, which was considered one of the herbicides least prone to such selection, has now been documented (25).

Given the increased dependence on proprietary herbicides and the corresponding herbicide-resistant genetically modified crops, the dissemination of such resistance to weeds represents a threat to both the environment and food supply. In particular, the impoverished, low plant diversity agroecosystems brought about through heavy use of herbicides favor competitive species adapted to the same broad-spectrum, post-emergence treatments that the herbicide-resistant crop strategy is designed to promote (65). Apart from the demonstrable production of herbicide-resistant weeds, it has been suggested that transgenic crops themselves may become weeds (63,64). The notion that crops and closely related weeds are necessarily separated from one another by numerous genetic differences is no longer accepted, and certain crops, such as strawberries, ryegrass, and broccoli, are actually weeds themselves in certain circumstances (reviewed in Rissler and Mellon, Reference 67). An even greater risk, however, is that large-scale releases of transgenic crops may promote transfer of transgenes from crops to other plants, which may then become weeds. The main mechanism for this would be hybridization between distinct plant species (17), although evidence has also accumulated for horizontal transfer of transgenes to soil bacteria (57), and concerns have been voiced in this regard about the widespread use of the highly recombinogenic cauliflower mosaic viral promoter in transgenic plants (30).

The example of teosinte, the wild grass from which corn is believed to have originated, is instructive with regard to potential transfer by hybridization. This plant grows in and around corn fields in Mexico and Guatemala, and produces full fertile hybrids with corn, with the weed freely exchanging genetic material with the crop under field conditions (20,81). There are also crops that are grown near wild/weedy plants that are not close relatives but may have some degree of cross compatibility (65). It is thus reasonable to expect that introduction of herbicide resistance genes into crops, consequent selection for herbicide resistance in wild plants, and subsequent exchanges with other wild plants, will eventually undermine the use of entire classes of herbicides for weed control.

The extensive use of chemical herbicides in agriculture has toxic effects on nontarget organisms, including animals and humans. Bromoxynil, an herbicide for which a genetically resistant variety of cotton was engineered, is absorbed through the skin and is teratogenic in mice and rats (9,68). It was banned by the EPA in 1998. Yet bromoxynil-resistant cotton has been approved in the U.S., and in Canada, where it is not grown but is approved for livestock feed.

Glyphosate is genotoxic by several in vitro (48) and in vivo (10) criteria. In addition, treatment of rabbits with sublethal doses of glyphosate resulted in a decline in body weight, libido, ejaculate volume, sperm concentration, semen fructose levels, and semen osmolality (88). Chemical herbicides thus represent a hazard to farm workers, and to the human population at large through the contamination of ground water. The magnitude of this hazard will be greatly exacerbated by increased use of herbicide-resistant crops.

The insertion of the gene encoding Bt toxin into crop plants to control insect pests is another use of gene modification fraught with hazards. Although it has been touted as a benign use of genetic technology that will permit the replacement of synthetic insecticides now used to control insect pests, most crops are susceptible to a diversity of such pests, and insecticides will still have to be used to control pests other than the Lepidoptera susceptible to Bt toxin (28). Moreover, several Lepidopteran species have been reported to develop resistance to Bt toxin in both field and laboratory tests, suggesting that major resistance problems are likely to develop in Bt crops, which will require moving to new chemical or gene-encoded pesticides (76).

An important factor in accelerating selection for resistance to Bt toxin in insects feeding on genetically modified plants is that unlike the natural exposure of insects at specific stages in their life cycles to bacteria living on exposed regions of plants during a limited portion of the growing season, insects will potentially be exposed to Bt toxin at all stages of their life cycles on all parts of the plant, for the entire growing season (54). The strategy of creating refuges around planted fields of Bt-resistant corn in order to inhibit evolution of resistance in target insects depends on the assumption that genetic resistance is a recessively inherited trait. However, recent work has indicated that, at least for the European corn borer *Ostrinia nubilalis*, this assumption is untrue (34).

Bt and other genetically introduced toxins also present a variety of hazards to nontarget organisms. For example, soil-inhabiting insects that degrade plant debris may be harmed by Bt toxin, disrupting natural recycling processes (62). A recent laboratory study found that larvae of the monarch butterfly, *Danaus plexippus*, reared on milkweed leaves dusted with pollen from Bt corn ate less, grew more slowly, and suffered higher mortality than larvae reared on leaves dusted with untransformed corn pollen or on leaves without pollen (49).

The potential for adverse effects of genetically modified crops on nontarget organisms raises the issue of the impact of foods produced in this fashion on human health. Speculations along these lines have been given credibility by a recent study in which rats were fed on potatoes genetically modified to express the snowdrop lectin *Galanthus nivalis* agglutinin (23), a variety developed to resist damage from insects and nematodes. The animals fed on the genetically modified potatoes exhibited pathological epithelial proliferation and elevated lymphocyte infiltration of digestive tract epithelia relative to rats fed on the parental strain of potato.

Allergies to proteins introduced by transgenesis is also a concern. A genetically modified variety of soybean containing the gene for the methionine-rich 2S albumin from the Brazil nut (*Betholletia excelsa*) was found to produce allergic reactions in subjects allergic to Brazil nuts (59). Since the effect of a given transgene on the expression of other genes during the development of a genetically modified plant is unknown and difficult to ascertain, it is likely that the concentrations of natural toxins (38) and other bioactive components of fruits and vegetables will be altered by transgenesis in an unpredictable fashion. Indeed, a recent study found substantial reductions in the levels of phytoestrogens, natural components of soybeans, in two different varieties that had been genetically modified for herbicide-resistance relative to the isogenic, nongenetically modified varieties (43).

Given all these uncertainties and demonstrable hazards, as well as the fact that no genetically modified foods to date are more nutritious or less expensive than their natural counterparts (77), it is not surprising that there have been calls from many organizations worldwide, including farmers, environmentalists, and health professionals, for a reevaluation of policies that permit the planting and marketing of transgenic crops (52). In our view, application of the precautionary principle would suggest that at the very least a moratorium on these activities is desirable to allow for further research and public discussion on the safety of, and need for, genetically modified crops. The rapid introduction of genetically modified crops and the millions of acres of such crops under cultivation suggests that synergistic effects leading to a cascade of problems affecting large portions of the food supply are possible, if not probable. In 1999, U.S. Department of Agriculture (USDA) spent under 1.5 million dollars on research into the ecological effects of biotechnology. This is a minuscule fraction of the amount of money any of the major agricultural biotechnology corporations is spending on developing new products. In fact, it is a minuscule fraction of the money spent by the federal government in developing these technologies. It has been estimated that during the past decade the USDA has invested over $2 billion in advancing biotechnology. The land grant system has invested probably another $1 billion in state funds (C. Benbrook, private communication). It is clear from the discussion above that these expenditures are out of balance, and that for precautionary measures to be actually implemented, the research dollars required to assess and control the hazards must come from the proponents of the technology, not the public.

Human Developmental Gene Modification

Genetic modification of human embryos or fetuses, referred to here as developmental modifications, has been proposed for purposes of both prevention of disease and enhancement of capacity. The development of sophisticated in vitro fertilization methods, preimplantation DNA analysis, improved techniques for gene transfer, insertion, or conversion, and embryo implantation procedures, have made such interventions technically feasible. The hazards of genetic modifications to humans have often been discussed in terms of somatic modification, in which only nonreproductive tissues are affected, and germline modification, in which changes to an individual's DNA can be passed down to future generations. Indeed, this division has led to the general belief that the only or main hazard of developmental modification is the potential of transmission of undesired alterations in the germline. But it is clear that the hazards of developmental gene modification are much more extensive.

The hazards of germline transmission of DNA modification are obvious from a reading of the literature on transgenic animals. For example, germline introduction of an improperly regulated normal gene resulted in progeny with unaffected development but enhanced tumor incidence during adult life (44). DNA modifications during embryogenesis may affect gene function in ways that are not immediately apparent, either perturbing development of the originally altered individual, or, if transmitted into the germline, that individual's progeny, in ways that may not be recognized for a generation or more. Interactions among genes and their products are highly integrated, have been refined over evolutionary time-scales, and often serve to stabilize developmental pathways and physiological homeostasis (56,75,82). In addition, cellular signaling pathways are overlapping, rather than discrete or separable (24) ensuring that any developmental genetic alteration will have broad, uncontrolled ramifications. Through experimental error, unanticipated allelic interactions, or poorly understood regulatory mechanisms such as imprinting, developmental genetic manipulation risks altering sensitive biological equilibria. Disrupting these interactive systems is likely to have complex and uncertain biological effects, including some which only become apparent during the development or functioning of specific cells or tissues (39).

Inadvertent developmental modification represents a developmental hazard of many of the more than 200 somatic gene therapy protocols now in use. For example, viral vectors employed in somatic gene therapy protocols can infect mouse eggs in vitro, leading to germline transmission of a transgene in the progeny (79). Although removal of the zona pellucida was a prerequisite for infection of the eggs in vitro, the early oocytes of postnatal ovaries also lack zonas. These experiments thus raise the possibility that modification of gametes may occur in vivo, with unknown and uncontrolled consequences.

In protocols that attempt somatic gene therapy for life-threatening illnesses, saving the life of the patient is a value that must be balanced against developmental risks, including those to the germline, and indeed, such considerations also pertain to chemotherapy in cancer patients (8). With respect to deliberate developmental modifications, the story is quite different. The unpredictability of embryo gene modification means that this procedure will inevitably be experimental when tried in humans, no matter how reliably it may be done in other species. No informed consent on the part of the potential parents is possible, because no reliable information would be available when it is first attempted. Performing such experiments on a developing human, which entails the possibility of experimenter-induced birth defects in that individual or its offspring, is scientifically and medically irresponsible and a violation of the health care practitioner's oath to do no harm. Here, there is no existing patient whose life is in jeopardy, but rather an embryo in a petri dish that the egg or sperm donor (or whoever else may gain the right to its disposition) would like to have turn out genetically differently. If ever there was a technology not yet in place, which the precautionary principle would warn us away from undertaking, developmental genetic modification of humans would seem to be it.

THE NEED FOR A PRECAUTIONARY SCIENCE

The preceding discussion and three case studies indicate why we need the precautionary principle: the potential harms are serious or irreversible and the uncertainties vast. The principle, in contrast to most decision rules used for policy, embeds ethics as well as epistemology. We are to act with precaution when the stakes are high and science will take a long time to link cause and effect.

As formulated by a group of scientists, lawyers, and nongovernmental organizations at the Wingspread Conference held in January 1998, the precautionary principle has four elements (66):

- People have a duty to take anticipatory action to prevent harm.
- The burden of proof of harmlessness of a new technology, process, activity, or chemical lies with the proponents, not with the public.
- Before using a new technology, process, or chemical, or starting a new activity, people have an obligation to examine "a full range of alternatives."
- Decisions applying the precautionary principle must be "open, informed, and democratic" and "must include affected parties."

Under current regulations, proponents must perform risk assessment experiments and submit data to governmental regulators. However, there is no requirement for independent review of these assessments, and several studies in the U.S. and Canada have suggested that risk assessments are inadequate to evaluate and predict complex ecological interactions. A freedom of information request to the Canadian Food Inspection Agency (the federal agency responsible for environmental safety of genetically modified crops) revealed that environmental assessments for the first commercial release of herbicide-tolerant canola were limited in the scope of questions asked, types of hazards considered, and the geographic and temporal scale over which the tests were conducted. Invasiveness of genetically modified crops into nonagricultural settings, to cite just one example, was assessed in a single experimental plot in a road-side ditch (Reference 4, and for U.S. examples, see also Reference 86).

Under these standard regulatory schemes, the burden of proof ultimately rests with the public to show that genetically modified microbes or crops are harmful. However, the public does not have access to sufficient information to analyze the accuracy of tests used or their validity, i.e., their relevance to real world problems. Extended to human developmental modifications as is now being proposed (74), these regulatory protocols threaten to make the experimentally altered individuals and their progeny the ultimate testing ground.

The commercial promotion of genetic modification technologies conceals viable alternatives to these potentially damaging activities. But U.S. law, the National Environmental Policy Act, requires alternatives to potentially harmful activities which are federally funded to be fully examined. It requires the least harmful methods to be used. Scientists can, and should, help assess alternatives to potentially destructive genetically modified organisms. But a precautionary science will also acknowledge that decision-making fraught with scientific uncertainties necessitates policy choices that scientists are no more qualified than lay people to make. Accordingly, the affected public must be intimately involved in deciding how and whether to proceed with genetic manipulation.

Acting with precaution will require scientists to pose different questions. It is no longer sufficient to ask if a technology is possible, but what possible damage might result from employing it in the real world. In cases where we decide to proceed with the technology, or it is too firmly entrenched to reverse, we will need to set up enough feedback loops so that alert practitioners in all disciplines and walks of life can discern systemic changes before it is too late. Farmers, mothers, physicians, and scientists need to have the information and skills to observe possible damage and report it. In particular, all genetically modified foods and crops that have been released need to be labeled so they can be monitored by consumers and experts.

As indicated by the cases discussed above, the precautionary principle will have very different applications depending on the severity of the potential harms and the uncertainties. The new products developed by genetic manipulation have far-outstripped the regulatory system's capacity to apply the requisite oversight on behalf of the public. The precautionary principle provides a new framework by which society can choose the technologies worth proceeding with and those which, for good reasons, should be held in abeyance or proscribed. Any genetic manipulation of living organisms that proceeds with disregard of these considerations cannot claim to be fully scientific.

SUMMARY

Technologies that genetically modify organisms have proved useful in the manufacturing of biologicals, have been used to alter food crops and farm animals, and have attracted interest as possible therapeutic modalities in humans. All of these actual and potential applications have engendered opposition variously based on safety and ethical and moral concerns. It was our purpose here to discuss a range of these technologies in terms of the "precautionary principle," an emerging international standard in decision making around implementation of technologies that complements in important ways traditional risk-benefit analysis. In particular, in the absence of compelling data on potential harms, the precautionary principle shifts the burden of proof from the opponents of implementation of a new technology to its advocates. In a formal sense, it inverts the commonplace among scientists that considers it "better science" to erroneously claim that there is no effect than to erroneously claim that there is an effect. We suggest that in its acknowledgement of the complexities of assessing the impact of technologies on the environment and public health, the precautionary principle represents a scientific improvement over standard risk-benefit analysis. We considered three genetic modification technologies as cases in which the application of the precautionary principle could lead to different conclusions about whether or not to proceed: *(i)* production of transgenic microorganisms, *(ii)* production of genetically modified food crops, and *(iii)* production of genetically modified human individuals.

REFERENCES

1.Altieri, M.A. 1998. The environmental risks of transgenic crops: an agroecological assessment (http://www.cnr.berkeley.edu/~agroeco3/risks.html).

2.Anderson, E.S. 1975. Viability of, and transfer of a plasmid from, *E. coli* K12 in human intestine. Nature *255*:502-504.

3.Atkinson, M.A. 1997. Molecular mimicry and the pathogenesis of insulin-dependent diabetes mellitus: still just an attractive hypothesis. Ann. Med. *29*:393-399.

4.Barrett, K. 1999. Canadian Agricultural Biotechnology: Risk Assessment and the Precautionary Principle. Department of Botany. University of British Columbia.

5.Barrett, K. and C. Raffensperger. 1999. Precautionary science, p. 106-122. *In* C. Raffensperger and J. Tickner (Eds.), Protecting Public Health and the Environment: Implementing the Precautionary Principle. Island Press, Washington, D.C.

6.Boehmer-Christiansen, S. 1994. The precautionary principle in Germany: enabling government, p. 31-68. *In* T. O'Riordan and J. Cameron (Eds.), Interpreting The Precautionary Principle. Earthscan Publications, London.

7.Bork, P. and R.F. Doolittle. 1992. Proposed acquisition of an animal protein domain by bacteria. Proc. Natl. Acad. Sci. USA *89*:8990-8994.

8.Bridges, B.A. 1979. An estimate of genetic risk from 8-methoxypsoralen photochemotherapy. Hum. Genet. *49*:91-96.

9.Chernoff, N., J.M. Rogers, C.I. Turner, and B.M. Francis. 1991. Significance of supernumerary ribs in rodent developmental toxicity studies: postnatal persistence in rats and mice. Fundam. Appl. Toxicol. *17*: 448-453.

10. **Clements, C., S. Ralph, and M. Petras.** 1997. Genotoxicity of select herbicides in Rana catesbeiana tadpoles using the alkaline single-cell gel DNA electrophoresis (comet) assay. Environ. Mol. Mutagen. *29*:277-288.

11. **Cohen, I.R.** 1991. Autoimmunity to chaperonins in the pathogenesis of arthritis and diabetes. Annu. Rev. Immunol. *9*:567-589.

12. **Cranor, C.F.** 1999. Asymmetric information, the precautionary principle, and burdens of proof, p. 106-122. *In* C. Raffensperger and J. Tickner (Eds.), Protecting Public Health and the Environment: Implementing the Precautionary Principle. Island Press, Washington, D.C.

13. **Cunningham, M.W.** 1993. Bacterial antigen mimicry, p. 245-261. *In* C. Bona, K. Siminovitch, M. Zanetti, and A. Theofilopoulos (Eds.), The Molecular Pathology of Autoimmune Diseases. Harwood Academic, Chur, Switzerland.

14. **Cunningham, M.W.** 1996. Streptococci and rheumatic fever, p. 13-66. *In* N.R. Rose and H. Friedman (Eds.), Microorganisms and Autoimmune Disease. Plenum Publishing, New York.

15. **Cunningham, M.W., S.M. Antone, M. Smart, R. Liu, and S. Kosanke.** 1997. Molecular analysis of human cardiac myosin-cross-reactive B- and T-cell epitopes of the group A streptococcal M5 protein. Infect. Immun. *65*:3913-3923.

16. **Dale, J.B. and E.H. Beachey.** 1985. Epitopes of streptococcal M proteins shared with cardiac myosin. J. Exp. Med. *162*:583-591.

17. **Darmency, H.** 1994. The impact of hybrids between genetically modified crop plants and their related species: introgression and weediness. Molec. Ecol. *3*:37-40.

18. **Davies, J.M.** 1997. Molecular mimicry: can epitope mimicry induce autoimmune disease? Immunol. Cell Biol. *75*:113-126.

19. **Dijkmans, R., F. Cornette, S. Kreps, E. Martens, J. Vankerkom, M. Mergeay, and A. Billiau.** 1993. Synergistic toxicity of IFN-gamma-producing *Escherichia coli* K12 cells. Microb. Releases *2*:23-28.

20. **Doebley, J.** 1990. Molecular evidence for gene flow among Zea species. Bioscience *40*:443-448.

21. **Eichbaum, Q.G., D.W. Beatty, and M.I. Parker.** 1994. Identification of cardiac autoantigens in human heart cDNA libraries using acute rheumatic fever sera. J. Autoimmun. *7*:243-261.

22. **Evans, C.F., M.S. Horwitz, M.V. Hobbs, and M.B. Oldstone.** 1996. Viral infection of transgenic mice expressing a viral protein in oligodendrocytes leads to chronic central nervous system autoimmune disease. J. Exp. Med. *184*:2371-2384.

23. **Ewen, S.W. and A. Pusztai.** 1999. Effect of diets containing genetically modified potatoes expressing Galanthus nivalis lectin on rat small intestine. Lancet *354*:1353-1354.

24. **Fambrough, D., K. McClure, A. Kazlauskas, and E.S. Lander.** 1999. Diverse signaling pathways activated by growth factor receptors induce broadly overlapping, rather than independent, sets of genes. Cell *97*:727-741.

25. **Gill, D.S.** 1995. Development of herbicide resistance in annual ryegrass populations in the cropping belt of Western Australia. Austral. J. Exp. Agric. *3*:67-72.

26. **Goldburg, R.J.** 1992. Environmental concerns with the development of herbicide-tolerant plants. Weed Technol. *6*:647-652.

27. **Gran, B., B. Hemmer, and R. Martin.** 1999. Molecular mimicry and multiple sclerosis—a possible role for degenerate T cell recognition in the induction of autoimmune responses. J. Neural. Transm. (Suppl) *55*:19-31.

28. **Gould, F.** 1994. Potential and problems with high-dose strategies for pesticidal engineered crops. Biocontrol Sci. Technol. *4*:451-461.

29. **Ho, M.-W., T. Traavik, O. Olsvik, B. Tappeser, C.V. Howard, C. von Weizsacker, and G.C. McGavin.** 1998. Gene technology and gene ecology of infectious diseases. Microb. Ecol. Health Dis. *10*:33-59.

30. **Ho, M.-W., A. Ryan, and J. Cummins.** 2000. Cauliflower mosaic viral promoter—a recipe for disaster? Microb. Ecol. Health Dis. *11*:194-197.

31. **Holt, J.S., S.B. Powles, and J.A.M. Holtum.** 1993. Mechanisms and agronomic aspects of herbicide resistance. Ann. Rev. Plant Phys. Plant Mol. Biol. *44*:203-229.

32. **Hooper, E.** 1999. The River: A Journey to the Source of HIV and AIDS. Little Brown, Boston.

33. **Hosein, B., M. McCarty, and V.A. Fischetti.** 1979. Amino acid sequence and physicochemical similarities between streptococcal M protein and mammalian tropomyosin. Proc. Natl. Acad. Sci. USA *76*:3765-3768.

34.Huang, F., L.L. Buschman, R.A. Higgins, and W.H. McGaughey. 1999. Inheritance of resistance to *bacillus thuringiensis* toxin (Dipel ES) in the European corn borer. Science *284*:965-967.

35.Iglewski, W.J. and S. Dewhurst. 1991. Cellular mono(ADP-ribosyl) transferase inhibits protein synthesis. FEBS Lett. *283*:235-238.

36.Iglewski, W.J., H. Lee, and P. Muller. 1984. ADP-ribosyltransferase from beef liver which ADP-ribosylates elongation factor-2. FEBS Lett. *173*:113-118.

37.Inageda, K., H. Nishina, and S. Tanuma. 1991. Mono-ADP-ribosylation of Gs by an eukaryotic arginine-specific ADP- ribosyltransferase stimulates the adenylate cyclase system. Biochem. Biophys. Res. Commun. *176*:1014-1019.

38.Jadhav, S.J., R.P. Sharma, and D.K. Salunkhe. 1981. Naturally occurring toxic alkaloids in foods. Crit. Rev. Toxicol. *9*:21-104.

39.Jaenisch, R., M. Breindl, K. Harbers, D. Jahner, and J. Lohler. 1985. Retroviruses and insertional mutagenesis. Cold Spring Harb. Symp. Quant. Biol. *50*:439-445.

40.Khanna, A.K., Y. Nomura, V.A. Fischetti, and J.B. Zabriskie. 1997. Antibodies in the sera of acute rheumatic fever patients bind to human cardiac tropomyosin. J. Autoimmun. *10*:99-106.

41.Krimsky, S. and R.P. Wrubel. 1996. Agricultural Biotechnology and the Environment. University of Illinois Press, Urbana.

42.Krisher, K. and M.W. Cunningham. 1985. Myosin: a link between streptococci and heart. Science *227*:413-415.

43.Lappé, M.A., E.B. Bailey, C. Childress, and K.D.R. Setchell. 1999. Alterations in clinically important phytoestrogens in genetically modified, herbicide-tolerant soybeans. J. Medicin. Food *1*:241-245.

44.Leder, A., P.K. Pattengale, A. Kuo, T.A. Stewart, and P. Leder. 1986. Consequences of widespread deregulation of the c-myc gene in transgenic mice: multiple neoplasms and normal development. Cell *45*:485-495.

45.Lee, H. and W.J. Iglewski. 1984. Cellular ADP-ribosyltransferase with the same mechanism of action as diphtheria toxin and *Pseudomonas* toxin A. Proc. Natl. Acad. Sci. USA *81*:2703-2707.

46.Levin, B.R. and R.V. Tauxe. 1996. Cholera: nice bacteria and bad viruses. Curr. Biol. *6*:1389-1391.

47.Levy, S.B., B. Marshall, D. Rowse-Eagle, and A. Onderdonk. 1980. Survival of *Escherichia coli* host-vector systems in the mammalian intestine. Science *209*:391-394.

48.Lioi, M.B., M.R. Scarfi, A. Santoro, R. Barbieri, O. Zeni, D. Di Berardino, and M.V. Ursini. 1998. Genotoxicity and oxidative stress induced by pesticide exposure in bovine lymphocyte cultures in vitro. Mutat. Res. *403*:13-20.

49.Losey, J.E., L.S. Rayor, and M.E. Carter. 1999. Transgenic pollen harms monarch larvae. Nature *399*:214.

50.Lubran, M.M. 1988. Bacterial toxins. Ann. Clin. Lab. Sci. *18*:58-71.

51.Manjula, B.N., B.L. Trus, and V.A. Fischetti. 1985. Presence of two distinct regions in the coiled-coil structure of the streptococcal Pep M5 protein: relationship to mammalian coiled-coil proteins and implications to its biological properties. Proc. Natl. Acad. Sci. USA *82*:1064-1068.

52.Marks, A. 1999. U.S. poised for a biotech food fight. Christian Science Monitor. November 17, p. 1.

53.Marx, A., A. Wilisch, A. Schultz, S. Gattenlohner, R. Nenninger, and H.K. Muller-Hermelink. 1997. Pathogenesis of myasthenia gravis. Virchows Arch. *430*:355-364.

54.McGaughey, W.H. and M.E. Whalon. 1992. Managing resistance to *Bacillus thuringiensis* toxins. Science *258*:1451-1455.

55.Newman, S.A. 1982. The "Scientific" selling of rDNA. Environment *24*:21-23; 53-57.

56.Newman, S.A. 1995. Interplay of genetics and physical processes of tissue morphogenesis in development and evolution: the biological fifth dimension, p. 3-12. *In* D. Beysens, G. Forgacs, and F. Gaill (Eds.), Interplay of Genetic and Physical Processes in the Development of Biological Form. World Scientific, Singapore.

57.Nielsen, K.M., A.M. Bones, K. Smalla, and J.D. van Elsas. 1998. Horizontal gene transfer from transgenic plants to terrestrial bacteria—a rare event? FEMS Microbiol. Rev. *22*:79-103.

58.NIH Guidelines, Appendix C (**http://www4.od.nih.gov/oba/apndxc.htm**).

59.Nordlee, J.A., S.L. Taylor, J.A. Townsend, L.A. Thomas, and R.K. Bush. 1996. Identification of a Brazil-nut allergen in transgenic soybeans. N. Engl. J. Med. *334*:688-692.

60.Okazaki, I.J., H.J. Kim, and J. Moss. 1996. Cloning and characterization of a novel membrane-associated lymphocyte NAD:arginine ADP-ribosyltransferase. J. Biol. Chem. *271*:22052-22057.

61. Oldstone, M.B. 1998. Molecular mimicry and immune-mediated diseases. FASEB J. *12*:1255-1265.

62. Palm, C., K. Donegan, D. Harris, and R. Seidler. 1994. Quantification in soil of *Bacillus thuringiensis* var. *kurstaki* delta-toxin from transgenic plants. Molec. Ecol. *3*:145-151.

63. Parker, I.M. and P. Kareiva. 1996. Assessing the risks of invasion for genetically engineered plants: acceptable evidence and reasonable doubt. Biol. Conserv. *78*:193-203.

64. Purrington, C.B. and J. Bergelson. 1995. Assessing weediness of transgenic crops: industry plays plant ecologist. Trends Ecol. Evol. *10*:340-342.

65. Radosevich, S.R., J.S. Holt, and C.M. Ghersa. 1996. Weed Ecology: Implications for Weed Management, 2nd ed. John Wiley & Sons, New York.

66. Raffensperger, C. and J. Tickner. (Eds.). 1999. Protecting Public Health and the Environment: Implementing the Precautionary Principle. Island Press, Washington, D.C.

67. Rissler, J. and M. Mellon. 1996. The Ecological Risks of Engineered Crops. MIT Press, Cambridge, MA.

68. Rogers, J.M., B.M. Francis, B.D. Barbee, and N. Chernoff. 1991. Developmental toxicity of bromoxynil in mice and rats. Fundam. Appl. Toxicol. *17*:442-447.

69. Sagik, B.P., C.A. Sorber, and B.E. Moore. 1981. The survival of EK1 and EK2 systems in sewage treatment plant models, p. 449-460. *In* S.B. Levy, R.C. Clowes, and E.L. Koenig (Eds.), Molecular Biology, Pathogenicity and Ecology of Bacterial Plasmids. Plenum Press, New York.

70. Schwimmbeck, P.L., T. Dyrberg, D.B. Drachman, and M.B. Oldstone. 1989. Molecular mimicry and myasthenia gravis. An autoantigenic site of the acetylcholine receptor alpha-subunit that has biologic activity and reacts immunochemically with herpes simplex virus. J. Clin. Invest. *84*:1174-1180.

71. Schwimmbeck, P.L. and M.B. Oldstone. 1989. Klebsiella pneumoniae and HLA B27-associated diseases of Reiter's syndrome and ankylosing spondylitis. Curr. Top. Microbiol. Immunol. *145*:45-56.

72. Sinsheimer, R. 1977. Mapping the mammalian genome: potential risks, p. 74-80; Discussion, p. 80-89. *In* Research with Recombinant DNA. National Academy of Sciences, Washington, D.C.

73. Spangler, B.D. 1992. Structure and function of cholera toxin and the related *Escherichia coli* heat-labile enterotoxin. Microbiol. Rev. *56*:622-647.

74. Stock, G.B. and J. Campbell. 2000. Engineering the Human Germline: An Exploration of the Science and Ethics of Altering the Genes We Pass to Our Children. Oxford University Press, Oxford.

75. Strohman, R.C. 1993. Ancient genomes, wise bodies, unhealthy people: limits of a genetic paradigm in biology and medicine. Perspect. Biol. Med. *37*:112-145.

76. Tabashnik, B.E. 1994. Genetics of resistance to *Bacillus thuringiensis*. Ann. Rev. Entomol. *39*:47-79.

77. Teitel, M. and K. Wilson. 1999. Genetically Engineered Food: Changing the Nature of Nature. Park Street Press, Rochester, VT.

78. Tian, J., P.V. Lehmann, and D.L. Kaufman. 1994. T cell cross-reactivity between coxsackievirus and glutamate decarboxylase is associated with a murine diabetes susceptibility allele. J. Exp. Med. *180*:1979-1984.

79. Tsukui, T., Y. Kanegae, I. Saito, and Y. Toyoda. 1996. Transgenesis by adenovirus-mediated gene transfer into mouse zona-free eggs. Nat. Biotechnol. *14*:982-985.

80. Waldor, M.K. and J.J. Mekalanos. 1996. Lysogenic conversion by a filamentous phage encoding cholera toxin. Science *272*:1910-1914.

81. Wilkes, H. 1977. Hybridization of maize and teosinte in mexico and guatamala and the improvement of maize. Econ. Botany *31*:254-293.

82. Wilkins, A.S. 1997. Canalization: a molecular genetic perspective. BioEssays *19*:257-262.

83. Williams, P.H. 1977. Plasmid transfer in the human alimentary tract. FEMS Microbiol. Lett. *2*:91-95.

84. Wreggett, K.A. 1986. Bacterial toxins and the role of ADP-ribosylation. J. Recept. Res. *6*:95-126.

85. Wright, S. 1994. Molecular Politics. Developing American and British Regulatory Policy for Genetic Engineering, 1972-1982. University of Chicago Press, Chicago.

86. Wrubel, R.P., S. Krimsky, and R.E. Wetzler. 1992. Field testing transgenic plants. Bioscience *42*:280-289.

87. Wucherpfennig, K.W. and J.L. Strominger. 1995. Molecular mimicry in T cell-mediated autoimmunity: viral peptides activate human T cell clones specific for myelin basic protein. Cell *80*:695-705.

88. Yousef, M.I., M.H. Salem, H.Z. Ibrahim, S. Helmi, M.A. Seehy, and K. Bertheussen. 1995. Toxic effects of carbofuran and glyphosate on semen characteristics in rabbits. J. Environ. Sci. Health B *30*:513-534.

INDEX